Lecture Notes in Networks and Systems

Volume 766

The series 'Lecture Notes in Networks and Systems" publishes the latest developments in Networks and Systems—quickly, informally and with high quality. Original research reported in proceedings and post-proceedings represents the core of LNNS.

Volumes published in LNNS embrace all aspects and subfields of, as well as new challenges in, Networks and Systems.

The series contains proceedings and edited volumes in systems and networks, spanning the areas of Cyber-Physical Systems, Autonomous Systems, Sensor Networks, Control Systems, Energy Systems, Automotive Systems, Biological Systems, Vehicular Networking and Connected Vehicles, Aerospace Systems, Automation, Manufacturing, Smart Grids, Nonlinear Systems, Power Systems, Robotics, Social Systems, Economic Systems and other. Of particular value to both the contributors and the readership are the short publication timeframe and the world-wide distribution and exposure which enable both a wide and rapid dissemination of research output.

The series covers the theory, applications, and perspectives on the state of the art and future developments relevant to systems and networks, decision making, control, complex processes and related areas, as embedded in the fields of interdisciplinary and applied sciences, engineering, computer science, physics, economics, social, and life sciences, as well as the paradigms and methodologies behind them.

Indexed by SCOPUS, INSPEC, WTI Frankfurt eG, zbMATH, SCImago.

All books published in the series are submitted for consideration in Web of Science.

For proposals from Asia please contact Aninda Bose (aninda.bose@springer.com).

Robert Burduk · Michał Choraś · Rafał Kozik ·
Paweł Ksieniewicz · Tomasz Marciniak ·
Paweł Trajdos
Editors

Progress on Pattern Classification, Image Processing and Communications

Proceedings of the CORES and IP&C Conferences 2023

Springer

Editors
Robert Burduk
Wrocław University of Science and Technology
Wrocław, Poland

Rafał Kozik
University of Technology and Life Science
Bydgoszcz, Poland

Tomasz Marciniak
University of Technology and Life Science
Bydgoszcz, Poland

Michał Choraś
University of Technology and Life Science
Bydgoszcz, Poland

Paweł Ksieniewicz
Wrocław University of Science and Technology
Wrocław, Poland

Paweł Trajdos
Wrocław University of Science and Technology
Wrocław, Poland

ISSN 2367-3370 ISSN 2367-3389 (electronic)
Lecture Notes in Networks and Systems
ISBN 978-3-031-41629-3 ISBN 978-3-031-41630-9 (eBook)
https://doi.org/10.1007/978-3-031-41630-9

This Springer imprint is published by the registered company Springer Nature Switzerland AG
The registered company address is: Gewerbestrasse 11, 6330 Cham, Switzerland

Paper in this product is recyclable.

Preface

This volume of AISC contains the proceedings of the multi-conference consisting of two known and prestigious conferences with long tradition, namely:

- The 13th International Conference on Image Processing and Communications (IP&C 2023).
- The 13th International Conference on Computer Recognition Systems (CORES 2023).

In 2023, we decided to hold a joint event in the city of Wrocław at University of Science and Technology. The event is co-organized by Bydgoszcz University of Science and Technology (PBS). The multi-conference is held virtually.

The multi-conference brought together the researchers, developers, practitioners, and educators in the fields of image processing, computer recognition systems, networks, security, and advanced computer systems.

The conference proceedings contains high-level papers which were selected through a strict review process. The presented papers are structured in several sections, namely:

- Advanced data analysis and machine learning.
- Image processing.
- Artificial intelligence: various applications and problems.

Without the high-quality submissions from the authors, the success of the conference would not be possible. Therefore, we would like to thank all authors and also the reviewers for the effort they put into their submissions and evaluation.

Wrocław, Poland
Bydgoszcz, Poland
Bydgoszcz, Poland
Wrocław, Poland
Bydgoszcz, Poland
Bydgoszcz, Poland

Robert Burduk
Michał Choraś
Rafał Kozik
Paweł Ksieniewicz
Tomasz Marciniak
Paweł Trajdos
Michał Woźniak
Conference Chairs

Organization

IP&C

Honorary Chairs

Ryszard Tadeusiewicz	AGH University of Krakow, Poland
Ryszard S. Choraś	Bydgoszcz University of Science and Technology, Poland

Chair

Michał Choraś	Bydgoszcz University of Science and Technology, Poland

CORES

Honorary Chair

Marek Kurzyński	Wroclaw University of Science and Technology, Poland

Chairs

Michał Woźniak	Wroclaw University of Science and Technology, Poland
Robert Burduk	Wroclaw University of Science and Technology, Poland

Contents

Image Processing

Artificial Intelligence: Various Applications and Problems

Advanced Data Analysis and Machine Learning

A Novel Approach of Multi-string Parsing for Syntactic Pattern Recognition

Mariusz Flasiński and Janusz Jurek

Abstract A novel approach to syntactic pattern recognition based on multi-string parsing is introduced in the paper. A methodological motivation of the research, which has resulted from application-oriented requirements, is presented. Formal foundations, an architectural model and a generic recognition algorithm of the approach are introduced. The approach allows one to recognize complex objects which should be represented by a series of images instead of one image. The series of images is represented with a sequence of aspectual patterns which are treated as a holistic aspect-integrated description of a recognized object.

1 Introduction

Syntactic pattern recognition, SPR, [2, 11, 13, 14] is one of the main groups of pattern recognition. Patterns are treated here as elements of a formal language, usually a string language. For such a language a formal (generative) grammar is defined as a generator of a set of patterns. Then, a syntax analyzer (parser) is constructed as a pattern recognizer/interpreter. If patterns to be analyzed by a computerized pattern recognition system are of structural nature rather than the numerical one, then such a methodology is more adequate and advantageous than the methodology based on (numerical) feature vectors recognized by standard classifiers. (For a discussion of this issue see e.g. [11].)

DPLL(k), grammars introduced in [7] are a classic SPR model [11]. A research into the development of this model has been conducted in two complementary directions for twenty years. Fundamental conceptual investigations have allowed us to develop necessary formal SPR tools like: extended *DPLL*(k) grammars and their

M. Flasiński (✉)
AGH University of Science and Technology, al. A. Mickiewicza 30, 30-059 Kraków, Poland
e-mail: mariuz.flasinski@uj.edu.pl

J. Jurek
Information Technology Systems Department, Jagiellonian University, ul. prof. St. Łojasiewicza 4, 30-348 Kraków, Poland

R. Burduk et al. (eds.), *Progress on Pattern Classification, Image Processing and Communications*, Lecture Notes in Networks and Systems 766,
https://doi.org/10.1007/978-3-031-41630-9_1

error-correcting parser for a recognition of distorted/vague patterns [9] and stochastic *DPLL*(k) grammars and their stochastic parser for a recognition of patterns from statistical samples [9]. The research into the enhancement of the generative power of *DPLL*(k) grammars and a more efficient *DPLL*(k) parser [16, 17] allowed us to improve the initial formal model.

The results obtained in this fundamental/theoretical direction of the research allowed us to conduct a research in the second application-oriented direction, resulting in a variety of applications of the *DPLL*(k) model, including: monitoring and controlling the complex high energy physics detector [7], Auditory Brainstem Response audiometry [8], electrical load forecasting [12], analysis of ultrasound images of fetal palates [18, 26].

During the user requirements analysis phase of the research into the last mentioned application area, an untypical, and unprecedented in the SPR area, requirement has appeared. According to physicians, which formulated the requirements for the SPR system, the need to define a sequence of aspectual patterns (instead of one pattern) for an object to be recognized was the crucial demand as compatible with medical diagnostic practice. Moreover, a computerized recognition process, which should be consistent with a diagnostic interpretation performed by physicians, should then operate on this sequence of patterns considered as one (holistic) aspects-integrated description of the object. Indeed, such a problem has not been considered in the area of syntactic pattern recognition till now (for a survey of SPR problems see, e.g. [11]).

To solve this problem, the original SPR method has been developed and the recognition system has been implemented and applied for the analysis of ultrasound images of fetal palates. Processing in the system, as in every SPR system [2, 11, 13, 14], is carried out in three phases: image (pre)processing, structural pattern generation and syntax analysis (parsing). The image (pre)processing was presented in [26], the structural pattern generation was introduced in [18] and the application for the analysis of ultrasound images together with the discussion of the experimental results were described in [18, 26]. The crucial, syntax analysis phase, based on the methodology of the multi-string parsing, has not been presented yet. This methodology is original and unique in the area of syntactic pattern recognition. Its presentation is the aim of the paper. Thus, the paper is of methodological/theoretical nature.

The main achievements presented in the paper include: a definition of formal concepts of a multi-string grammar/language model, a development of an architectural model of a multi-string *DPLL*(k) syntax analyzer for syntactic pattern recognition and a construction of an algorithm of multi-string *DPLL*(k) parsing. They are presented in the succeeding sections.

2 Methodological Motivation

From the methodological point of view, a development of a syntactic pattern recognition model proceeds usually according to the following scheme [11]. Firstly, a generic method of the model, which is based on the standard form (in the sense considered in the theory of formal languages) of a grammar generating "pure" languages, is constructed for an interpretation and recognition of unambiguous patterns. Then, if the model is to be applied to application areas, in which patterns are usually distorted/vague, enhanced methods are constructed.

Generally, there are two causes of distortions and vagueness of structural patterns [11]. First of all, pattern classes can be defined on the basis of statistical samples of structural patterns, which can additionally be blurred and indistinct. In this case, stochastic [24, 25, 27] or fuzzy (in the sense of fuzzy set theory) [1, 5, 21] grammars are defined and the corresponding parsers are constructed as syntactic pattern recognizers. Secondly, errors which have arisen at an image (pre)processing phase can result in a generation of distorted structural representations. The use of the so-called extended grammars (containing also productions which represent such distortions) and the corresponding error-correcting parsers helps us to neutralize these unwanted effects occurred during image (pre)processing [3, 4, 15, 20].

As we have mentioned in Introduction, our *DPLL*(k) grammar-based model of syntactic pattern recognition has gone through all above mentioned stages of development, i.e. the generic method based on standard ("pure") *DPLL*(k) grammars has been proposed [7], its enhanced version based on stochastic *DPLL*(k) grammars and their stochastic parser, used for a recognition of patterns from statistical samples and for a recognition of blurred/indistinct patterns, has been constructed in [9] as well as an enhanced version based on extended *DPLL*(k) grammars and their error-correcting parser [9], which has been constructed in order to handle the problem of errors arisen at an image (pre)processing phase. However, during an application-oriented research into the use of *DPLL*(k) grammars for analysis of ultrasound images of fetal palates [18, 26], an unprecedented in the SPR area requirement has appeared.

The research has concerned the automatic diagnostics of congenital defects of fetuses, especially cleft palate. An analysis of images showing palatal bone structures has turned out to be difficult because of both small fetal structures (the overall size of fetuses diagnosed has varied from 45 to 84 mm) and variable echogenicity within a bone of a palate. The analysis of a *single image* of the shape of the palate bone has not been sufficient to assess the continuity of the structure, which has been the goal of the diagnosis. In fact, physicians do not analyze one "optimum" image[1] in this case, but the *sequence of images* of palatal bone structures, each of them with various binarization thresholds. (In our case there have been four images.) Of course, each of such images is represented in our approach with an adequate string. So, *a multi-string pattern* is to be analyzed in order either to ascertain palatal bone structure continuity or to assess a degree of structure loss that is further used to diagnose fetal palatal abnormalities.

In order to operate on a sequence of strings/patterns considered as one (holistic) aspects-integrated description of an object, a novel methodology of syntactic pattern recognition had to be developed. This requirement has motivated us into a research

---

[1] The "optimum" image would mean the image with the "optimum" value of the binarization threshold. Simply, there is no such an "optimum" threshold in this case.

into a multi-string *DPLL*(k) grammar-based model of syntactic pattern recognition. The results of this research are presented in the succeeding sections.

3 Preliminaries

Basic definitions which allow us to present the multi-string *DPLL*(k) parsing model as a pattern recognition scheme are presented in this section. The *programmed grammars*, introduced by Rosenkrantz in 1969 [22], are a classical model of increasing a generative power of the Chomsky context-free grammars, *CFGs*. This increase is obtained by a control of a derivation process with the help of fields containing indices of the productions to be applied. These fields can be ascribed to each *CFG* production. In the Rosenkrantz model the fields have to be pre-defined when the grammar is defined. (So, they are *static*.) The concept of *dynamically programmed grammars* allowing one to define a *CFG* grammar with fields, which can be filled *dynamically* during a derivation process, was introduced in 1995 [11]. Let us introduce the following definition.

Definition 1
A *dynamically programmed* (context-free), *DP*, grammar is a quintuple
$G = (\Sigma_N, \Sigma_T, P, S)$, where
Σ_N is a set of nonterminal symbols,
Σ_T is a set of terminal symbols,
P is a set of n productions of the form:
$p_i = (\pi_i, L_i, R_i, A_i, DCL_i)$, in which
$\pi_i : \bigcup_{k=1,\ldots,n} DCL_k \rightarrow \{TRUE, FALSE\}$ is the predicate of applicability of p_i,
$L_i \in \Sigma_N$ and $R_i \in \Sigma^*$ are the left- and right-hand sides of p_i, respectively,
A_i is the sequence of actions of a type: *add, read, move* which are performed over $\bigcup_{k=1,\ldots,n} DCL_k$,
DCL_i is the derivation control tape for p_i,
S is the start symbol (axiom), $S \in \Sigma_N$. □

(L_i, R_i) is called the core of p_i. For each $p_i, p_j \in P$, $i \neq j$, the core of p_i is different from the core of p_j. The operation $add(k, m)$ writes the symbol m on the cell of DCL_k which is under the head. The operation $read(k)$ returns the value read by the head of DCL_k. The operation $move(k)$ moves the head of DCL_k right.

In order to construct a polynomial parsing algorithm, two concepts of: *DP* grammars and well-known parsable *LL*(k) grammars [19, 23] have been used to define *DPLL*(k) grammars in [7]. Now, we introduce the following denotations and definitions.

Let $G = (\Sigma_N, \Sigma_T, P, S)$ be a *DP* grammar, $\eta \in \Sigma^*$, and $\mid x \mid$ denote the length of a string $x \in \Sigma^*$. $FIRST_k(\eta)$ denotes a set of all the terminal prefixes of the strings of the length k (or of a length less than k, if a terminal string shorter than k is derived from η) that can be derived from η in the grammar G that can be formalized in the following way.

$$FIRST_k(\eta) = \{x \in \Sigma_T^* : (\eta \overset{*}{\Longrightarrow} x\beta \wedge \mid x \mid = k) \\ \vee (\eta \overset{*}{\Longrightarrow} x \wedge \mid x \mid < k) \ , \ \beta \in \Sigma^*\}.$$

Definition 2
Let $G = (\Sigma_N, \Sigma_T, P, S)$ be a *DP* grammar, $\underset{\text{core}}{\overset{*}{\Longrightarrow}}$ denotes a sequence of derivation steps consisting in the applying of production cores only. G is called a *DPLL(k) grammar* iff the following two conditions are fulfilled.

1. For every two leftmost derivations in G

$$S \overset{*}{\Longrightarrow} \alpha A\theta \Longrightarrow \alpha\beta\theta \underset{\text{core}}{\overset{*}{\Longrightarrow}} \alpha x$$

$$S \overset{*}{\Longrightarrow} \alpha A\theta \Longrightarrow \alpha\gamma\theta \underset{\text{core}}{\overset{*}{\Longrightarrow}} \alpha y,$$

   where $\alpha, x, y \in \Sigma_T^*, \ \beta, \gamma, \theta \in \Sigma^*, \ A \in \Sigma_N,$ the following condition holds.

$$\text{If } FIRST_k(x) = FIRST_k(y), \ \text{ then } \beta = \gamma.$$

2. For G there exists $m > 0$ such that for any leftmost derivation $S \overset{*}{\Longrightarrow} \alpha A\theta \overset{\sigma}{\Longrightarrow} \alpha\beta\theta$, where σ is the string of the indices of productions applied, if $\mid \sigma \mid \geq m$ then the first symbol of $\beta\theta$ is terminal. □

As we have mentioned in Introduction, *DPLL(k)* grammars have the big generative power. They generate, apart from all the languages generated by *CFGs*, the considerable subclass of languages generated by the Chomsky *CSGs*, like $\{a^n b^n c^n : n \geq 0\}$, $\{a^n b^m c^n d^m : n, m \geq 0\}$, $\{a^{2^n} : n \geq 0\}$. The improved *DPLL(k)* parser of the $\mathcal{O}(n)$ time complexity was presented in [16]. Their practical applications have been listed in Introduction. *DPLL(k)* grammars will be used as a basis for constructing a multi-string parser in the succeeding section.

4 Multi-string Parsing

Before presenting the model of multi-string *DPLL(k)* parsing and corresponding algorithms, we introduce the following formal definitions concerning the concepts of multi-string *DPLL(k)* languages.

Definition 3
Let Σ_T be a set of terminal symbols. *A multi-string pattern m* is a sequence of d words: $m = (m_1,..., m_d)$, where $m_j \in \Sigma_T^*, \ j = 1, \ldots, d$ is called the *component* of the multi-string pattern m, d is called the *dimension* of m. □

Definition 4
A *multi-string DPLL(k) grammar G* is a sequence of grammars $G = (G^1, ..., G^d)$ where $G^j = (\Sigma_N^j, \Sigma_T, P^j, S^j),\ j = 1, \ldots, d$ is a *DPLL*(k) grammar. d is called the *dimension* of the multi-string grammar G. □

Let us notice that all grammars in a multi-string grammar share the same set of terminal symbols but they can have various: sets of non-terminal symbols, sets of productions, and the starting symbols.

Definition 5. Let $G = (G^1, ..., G^d)$ be a multi-string *DPLL*(k) grammar. A sequence of sets $L(G) = L(G^1), ..., L(G^d)$ is called a *multi-language generated with a DPLL(k) grammar G*. □

As it can be seen in Definition 2, the *DPLL*(k) parser [7] makes predictions during syntax analysis, looking ahead to the successive k-length prefixes of the input word. Let us formalize the notion of the *i-th k-prefix* of a word as follows.

Definition 6. Let $a = a_1 a_2 \ldots a_n \in \Sigma_T^*$ be a word. The prefix of a which begins with a_i, $i = 1, \ldots, n$, and it is of the length k, i.e. $a_i a_{(i+1)} \ldots a_{(i+k-1)}$, (or it contains all the symbols of a from a_i till a_n if its length is less than k) is called the *i-th k-prefix* of a.

Now, we can present the model of multi-string *DPLL*(k) parsing. Its scheme is shown in Fig. 1. A d-dimensional multi-string pattern is read into the input, which consists of d input tapes. The input head indicates the place i, which a symbol with its k-prefix is read from. The **Scheduler Module** is responsible for a delivery of the current k-prefix, denoted kpref[i,j], to a proper **Parser[j] Module**, j $= 1, \ldots, d$.

The main part of the system consists of a set of d *DPLL*(k) parsers. Each **Parser[j] Module** is constructed on the basis of the grammar G^j which generates the language $L(G^j)$ describing the j-th components, j $= 1, \ldots, d$, of multi-string patterns. So, any parser *differs* from the others, i.e. every parser has its specific control table. The indices of productions applied by a given parser during syntax analysis (which identify the analyzed component), as well as the parsing result, which are denoted together as: aspectual result[j] in Fig. 1, are both sent to the working memory, called **aspectual results**.

The **Aspectual Derivations Integrator** is responsible for the integration of the aspectual results obtained during the multi-string *DPLL*(k) parsing and for the delivering of the final holistic result. In our model we have assumed the criterion which is widely used in the area of AI for the conflict resolution in rule-based systems [10]. In one of the most popular method, namely the *highest priority rule* method, a priority corresponding to an importance determined by a human expert (e.g. a physician, a lawyer etc.) is ascribed to each rule. Such an approach, that is the use of priorities specified by the expert for components of multi-string pattern, has been applied in the model of multi-string *DPLL*(k) parsing for the implementation of the **Aspectual Derivations Integrator**.

Let us specify the model of multi-string *DPLL*(k) parsing in a more detailed way with a pseudocode. Firstly, we define its variables, data structures and functions in the following way.

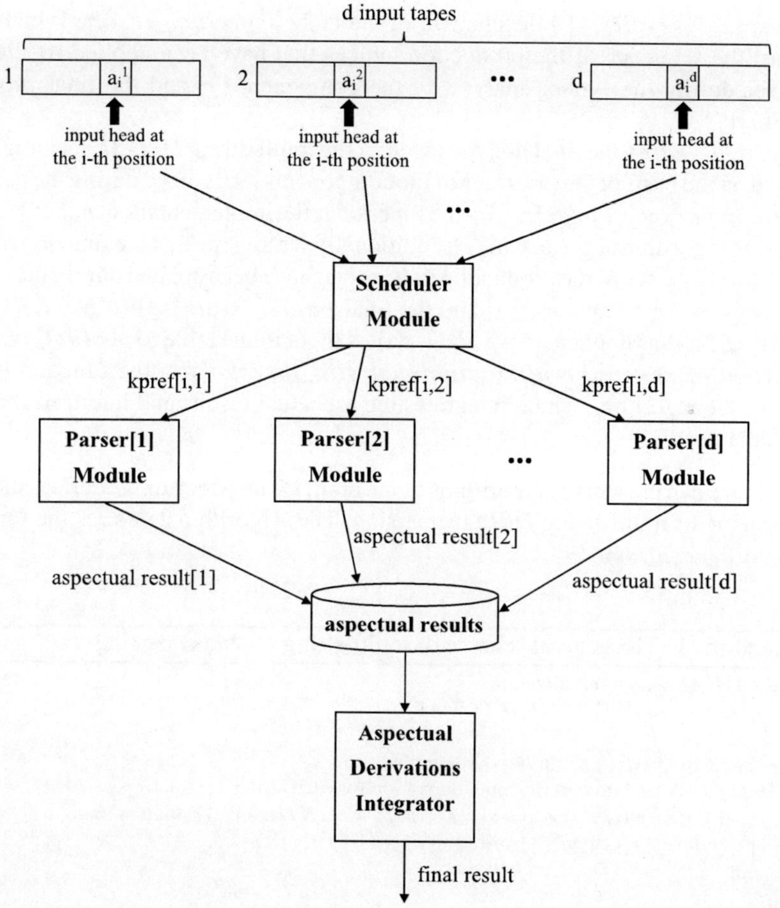

Fig. 1 The multi-string DPLL(k) parsing scheme.

- $length[j]$ - the length of the component $[j]$, $j = 1, \dots, d$, of the input multi-string pattern,
- $maxlength$ - is computed as $max\{length[j], j = 1, \dots, d\}$,
- $kpref[i, j]$ - the i-th k-prefix of the component $[j]$,
- $parser\_status[j]$ - the status of the j-th parser described as: $UNDERWAY$, ACC, ERR, where ACC means that the j-th component has been accepted, ERR - the j-th component has been rejected, $UNDERWAY$ - the parsing process is in progress,
- $actual\_holistic\_threshold$ - the threshold used for the final overall acceptance,
- $component\_priority[j]$ - the priority of the component $[j]$ in the multi-string pattern (pre)defined on the basis of the importance of the component determined by a (human) expert,

- $aspectual\_results[j]$ - the element of the table $aspectual\_results$ which contains: the sequence of the production indices that have been applied by the j-th parser during the syntax analysis of the component $[j]$ and the final $parser\_status[j]$,
- $approval\_rate$ - the final holistic result of the multi-string *DPLL*(k) parsing computed on the basis of the priorities of the components recognized during the parsing,
- *function* $parse(j, kpref[i, j])$ - the function is the implementation of the *DPLL*(k) parsing algorithm described in [7]; additionally we assume that the function returns both the sequence of the production indices that have been applied during the syntax analysis of the component $[j]$ and the final $parser\_status[j]$ ($UNDERWAY$, ACC, ERR) as defined above and it writes them to the table $aspectual\_results$,
- *function* $integrate\_results(actual\_holistic\_threshold)$ - the function implements the algorithm, which integrates the aspectual results and it returns the final holistic result.

Now, we can present the algorithms of the model. The Algorithm 1 defines the general scheme of multi-string *DPLL*(k) parsing. The Algorithm 2 defines the function $integrate\_results()$.

Algorithm 1: The general scheme of multi-string *DPLL*(k) parsing.

```
for j ← 1 to d do // initialization
    parser_status[j] ← UNDERWAY;
end
for i ← 1 to maxlength do // moving input head
    for j ← 1 to d do // performing derivation for each string
        if i <= length[j] ∧ parser_status[j] = UNDERWAY then
            parser_status[j] ← parse(j, kpref[i, j]);
    end
end
actual_holistic_threshold ← determine_threshold(aspectual_results);
write(integrate_results(actual_holistic_threshold)); // final result
for j ← to d do // additional information
    write(aspectual_results[j]);
end
```

The concepts of multi-string *DPLL*(k) parsing presented in this section and stochastic *DPLL*(k) grammars [9] discussed in the second section could be incorporated into one hybrid model for recognizing multi-aspectual distorted/vague patterns. A development of such a hybrid syntactic pattern recognition scheme would require a model that takes into account a combination of parameters of various meaning, i.e. the probability measure used for an assessment of a degree of pattern distortion/vagueness and the (subjective) priority determined by a human expert which corresponds to an importance of a given aspect.

Algorithm 2: The definition of the function *integrate_results*().

```
Function integrate_results(actual_holistic_threshold);
approval_rate ← 0;
for j ← 1 to d do
    if parser_status[j] = ACC then
        approval_rate ← approval_rate + component_priority[j] ;
end
if approval_rate > actual_holistic_threshold then return ACCEPT;
else return ERROR;
```

5 Concluding Remarks

The novel approach which extends the methodological paradigms of syntactic pattern recognition has been presented in the paper. This unique paradigmatic model is based on the idea of representing a pattern by a multi-aspectual structural description. Since as of yet, we have introduced this model for string languages, we are dealing with multi-string patterns. Therefore, the multi-string parsing model, which delivers the holistic aspects-integrated final result, had to be constructed. The efficient *DPLL*(*k*) parsing scheme [7] has been used as a basis for the construction of this multi-string parsing model. We have presented only the parsing (syntax analysis) phase in this paper, since preceding phases of processing in the SPR system, i.e. image (pre)processing and structural pattern generation as well as the application for the analysis of ultrasound images of fetal palates, have been described elsewhere ([18, 26]).

The model proposed will be further developed in two directions. Firstly, the model will be extended to the hybrid model for recognizing multi-aspectual distorted/vague patterns that has been mentioned at the end of the previous section. Secondly, the model will be extended for more complex patterns which are represented by graph structures. The efficient *ETPL*(*k*) graph grammar-based parsing scheme was proposed in [6] for syntactic pattern recognition and it was used in a variety of application areas, including scene analysis, Computer Aided Design/Computer Aided Manufacturing, analysis of visual events and activities, process monitoring and control (for a survey of its applications see [11]). Therefore, this scheme seems to be a good candidate to be extended to the multi-graph parsing model.

References

1. Asveld, P.R.J.: Fuzzy context-free languages (Part I and II). Theor. Comput. Sci. **347**, 167–213 (2005)
2. Bunke, H., Sanfeliu, A. (eds.): Syntactic and Structural Pattern Recognition - Theory and Applications. World Scientific, Singapore (1990)
3. Chan, K.F., Yeung, D.Y.: Error detection, error correction and performance evaluation in on-line mathematical expression recognition. Pattern Recogn. **34**, 1671–1684 (2001)
4. Cruz-Alcazar, P.P., Vidal, E.: Two grammatical inference applications in music processing. Appl. Artif. Intell. **22**, 53–76 (2008)
5. Ferreira, M., Santos, C., Monteiro, J.: Cork parquet quality control vision system based on texture segmentation and fuzzy grammar. IEEE Trans. Ind. Electron. **56**, 756–765 (2009)
6. Flasiński, M.: On the parsing of deterministic graph languages for syntactic pattern recognition. Pattern Recogn. **26**, 1–16 (1993)
7. Flasiński, M., Jurek, J.: Dynamically programmed automata for quasi contexts sensitive languages as a tool for inference support in pattern recognition-based real-time control expert systems. Pattern Recogn. **32**, 671–690 (1999)
8. Flasiński, M., Reroń, E., Jurek, J., Wójtowicz, P., Atłasiewicz, K.: On the construction of the syntactic pattern recognition-based expert system for Auditory Brainstem Response analysis. In: Kurzyński, M., Puchała, E., Woźniak, M., Żołnierek, A. (eds.) CORES 2005. Advances in Soft Computing, vol. 30, pp. 503–510, Springer, Heidelberg (2005). https://doi.org/10.1007/3-540-32390-2_59
9. Flasiński, M., Jurek, J.: On the analysis of fuzzy string patterns with the help of extended and stochastic GDPLL(k) grammars. Fundamenta Informaticae **71**, 1–14 (2006)
10. Flasiński, M.: Introduction to Artificial Intelligence. Springer, Switzerland (2016)
11. Flasiński, M.: Syntactic Pattern Recognition. World Scientific, Singapore (2019)
12. Flasiński, M., Jurek, J., Peszek, T.: On the learning of vague languages for syntactic pattern recognition. Pattern Anal. Appl. (2022). https://doi.org/10.1007/s10044-022-01120-0
13. Fu, K.S.: Syntactic Pattern Recognition and Applications. Prentice Hall, Upper Saddle River (1982)
14. Gonzales, R.C., Thomason, M.G.: Syntactic Pattern Recognition: An Introduction. Addison-Wesley, Boston (1978)
15. Huang, K.Y.: Syntactic Pattern Recognition for Seismic Oil Exploration. World Scientific, Singapore (2002)
16. Jurek, J.: On the linear computational complexity of the parser for quasi context sensitive languages. Pattern Recogn. Lett. **21**, 179–187 (2000)
17. Jurek, J.: Recent developments of the syntactic pattern recognition model based on quasi-context sensitive languages. Pattern Recogn. Lett. **26**, 1011–1018 (2005)
18. Jurek, J., Wójtowicz, W., Wójtowicz, A.: Syntactic pattern recognition-based diagnostics of fetal palates. Pattern Recogn. Lett. **133**, 144–150 (2020)
19. Lewis, P.M., II., Stearns, R.E.: Syntax-directed transduction. J. ACM **15**, 465–488 (1968)
20. Lopez, D., Sempere, J.M., Garcia, P.: Error correcting analysis for tree languages. Int. J. Pattern Recogn. Artif. Intell. **14**, 357–368 (2000)
21. Mordeson, J.N., Malik, D.S.: Fuzzy Automata and Languages: Theory and Applications. Chapman and Hall - CRC, Boca Raton (2002)
22. Rosenkrantz, D.J.: Programmed grammars and classes of formal languages. J. ACM **16**, 107–131 (1969)
23. Rosenkrantz, D.J., Stearns, R.E.: Properties of deterministic top-down grammars. Inf. Control **17**, 226–256 (1970)
24. Thomason, M.G.: Generating functions for stochastic context-free grammars. Int. J. Pattern Recogn. Artif. Intell. **4**, 553–572 (1990)
25. Vidal, E., Thollard, F., Higuera, C., Casacuberta, F., Carrasco, R.: Probabilistic finite-state machines (Part I and II). IEEE Trans. Pattern Anal. Mach. Intell. **27**, 1013–1039 (2005)

26. Wójtowicz, A., Wójtowicz, W., Jurek, J., Huras, H.: Evaluation of the fetal palate at 11 to 13 (+6) weeks of gestation based on an analysis of static ultrasound images using modern IT techniques. Prenatal Diagn. **38**, 414–421 (2018)
27. Zhu, S.C., Mumford, D.: A stochastic grammar of images. Found. Trends Comput. Graph. Vision **2**, 259–362 (2007)

Hollow n-grams Vectorizer for Natural Language Processing Problems

Weronika Borek-Marciniec and Paweł Ksieniewicz

Abstract Words don't come easy, which fosters the use of generative artificial intelligence models in ongoing popularity of widely available applications such as ChatGPT. The result is an even greater flood of online content that takes time to process. It is where Natural Language Processing tools for classification come in handy. Distinguishing fake news, event types, and other tasks can help process everyday information. In practice, such systems must work on data streams where fast prediction is needed. To achieve it, methods not based on neural networks can be used. Instead, they require feature extractions from the text to convert it to the model input. The primary methods used for this purpose are bag-of-words and n-grams, which allow converting the corpus of texts into a numerical format. This paper proposes a new strategy for creating n-grams – *Hollow n-grams* – which can be used to create classifiers ensembles with a higher generalization ability than models based on regular n-grams only.

Keywords machine learning · Natural Language Processing · n-grams

1 Introduction

Every day, humankind produces more data that floods us in everyday life, especially when using the Internet. It includes advertisements, audio or video materials, photos, messages, or regular content overflowing through social media. The use of artificial intelligence to organize this content or select materials recommended for individual users is becoming more common daily and goes beyond the scope of simple recommendation systems [12].

Recently, the GPT-3 model (*Generative Pre-trained Transformer*) has received extremal popularity, and soon the same will probably happen to its successor,

W. Borek-Marciniec (✉) · P. Ksieniewicz
Department of Systems and Computer Networks, Wroclaw University of Science and Technology, Wyb. Wyspiańskiego 27, 50370 Wroclaw, Poland
e-mail: weronika.borek@pwr.edu.pl

R. Burduk et al. (eds.), *Progress on Pattern Classification, Image Processing and Communications*, Lecture Notes in Networks and Systems 766,
https://doi.org/10.1007/978-3-031-41630-9_2

GPT-4, which is supposed to deal not only with text but also with multimodal data, e.g., correlated input of texts and images. Generating answers to questions, content on a specific topic, or descriptions of images, useful, for example, in creating audio descriptions for the blind, is just one of the challenges facing the *Natural Language Processing* field.

GPT-3 chat lacks the ability to verify quoted information, or colloquially speaking, "to tell the truth", which users often do not realize. Hence the problems emerging in schools and universities, such as using chat to write essays, reports, or literature reviews with imaginary references [5]. The apparent universality of the tool and its commonness meant that users, exposed to this type of artificial intelligence for the first time, were delighted with its advantages when the lack of knowledge about its operating results in the use of generated content in an inappropriate or even incorrect way [3]. It is just another proof that with the current level of technology development and the speed of content dissemination, people cannot verify the truth of the information they receive independently.

Applications that use NLP in classification tasks come to the rescue here – verifying the correctness of content [4], detecting fake news [10], classifying events [15] or sentiments [8].

Natural Language Processing methods not based on deep neural networks are often used in these solutions. Such an approach allows for shortening the learning time and response of the model, which is particularly useful in stream processing – necessary for systems used in practice with real data [11]. In addition, due to the lower complexity of this group of methods than deep learning, they consume less available resources, which contributes to the reduction of energy consumption and hence – minimization of the carbon footprint [7].

However, non-deep learning methods need to use preprocessing to convert non-serialized content to vector form. The most basic extraction method is bag-of-words, which consists of counting the occurrences of individual words in the text. Its more complex modification is based on n-grams, which use combinations of two or more words instead of single words. N-grams allow for the preservation of information on the connection of individual phrases, which may result in an increase in information density and, consequently, better generalization abilities of the model.

The building of n-grams, regardless of the n value, is also based on calculating the weights of specific words. Two basic approaches to this processing step are emphasized in the literature: *Term Freguency* (TF) and *Term Frequency – Inverse Document Frequency* (TF- IDF) [1]. The first assigns the highest weight to words that appear most often, which does not always result in the best quality because it may lead to favoring phrases popular for a given group of texts which appear in each of the analyzed documents. Hence, the use of the IDF strategy considers the percentage of texts in the entire corpus in which a given phrase appears and reduces its weight accordingly.

In this paper, we focus on text preprocessing and propose a new strategy for building a set of n-grams for $n > 1$, which can strengthen the classification quality by using it to create a heterogeneous set of estimators. Many studies show that when properly applied, integrating classifiers or classifier outputs positively affects the model's generalization abilities and prediction certainty. The reasons for this phenomenon were speculated as early as 1906 by Galton, who observed them at a cattle market. Classifier ensembles are still popular solutions in the field of machine learning thanks to its positive influence on methods quality [14].

The main contributions of this work are:

- proposition of the new strategy of building n-grams – *Hollow n-grams*,
- complete experiments checking its quality in compare to state-of-the-art NLP preprocessing methods,
- proposition of the classifiers ensemble creation with the usage of *Hollow n-grams*.

2 Hollow n-grams

The *Hollow n-grams* method is based on building n-grams at $n > 1$. However, instead of accomplishing this task by creating strings of words from adjacent words, it omits the word immediately following the analyzed starting point, as illustrated by Table 1.

The motivation for this solution was language construction, which puts directly related words – for example, adjectives or verbs for a specific noun – next to each other. It has obvious advantages but also leads to the omission of essential correlations between words more distant, which is perfectly visible for simple phrases where two nouns are separated by a verb – "**cook** makes **soup**" or "**student** starts **class**".

Therefore, the proposed method is not a stand-alone preprocessing process for classification tasks in the field NLP due to omitting the closest link between words. However, it can work as a supplement to information to be used in classifiers ensembles.

Table 1 Comparision of the regular n-grams with proposed strategy

n-gram	Sentence	Regular n-grams	Hollow n-grams
2	The quick brown fox jumps over the lazy dog.	"The quick", "quick brown", "brown fox", "fox jumps", "jumps over", "over the", "the lazy", "lazy dog".	"The brown", "quick fox", "brown jumps", "fox over", "jumps the", "over lazy", "the dog".
3	Real stupidity beats artificial intelligence every time.[1]	"Real stupidity beats", "stupidity beats artificial", "beats artificial intelligence", "artificial intelligence every", "intelligence every time"	"Real beats artificial", "stupidity artificial intelligence", "beats intelligence every", "artificial every time"

[1] Terry Pratchett, *Hogfather*

3 Experiments Design

In the experiments, the language *Python* and a library dedicated to machine learning – *scikit-learn* [13] were used. We picked *balanced accuracy score* [9] as a metric to offset the impact of imbalance on the results and cross-validation 5×2 as the experimental protocol to ensure results stability.

Multi-layer peceptron (MLP) was used as a model of the simplest neural network for a classification task. The classifier's parameters are set as:

- one layer with 100 neurons,
- ReLu activation function,
- 200 iterations maximum,
- "adam" solver.

E1 In the first experiment, the quality of the proposed method is compared with (a) the regular way of creating n-grams and (b) the classifier ensemble after integrating both approaches. TF- IDF and bigrams (n-grams for $n = 2$) were used to create the corpus. The experiment was carried out for different thresholds of the maximum number of final features – 50, 100, 200, 300, 400, and 500.

E2 The second experiment compares the accuracy of *Hollow n-grams*, regular n-grams, and a classifiers ensemble for a maximum number of features equal to 500 over different values of n – 2, 3, and 4. The study aims to compare differences in the quality of all approaches.

The dataset NELA-GT-2020 [6] was used in the research. The full version contains 1, 779, 127 articles in English from 529 sources. Article labels deal with the problem of detecting fake news – *reliable* texts are marked as 0, *mixed* – as class 1, and *unreliable* – as class 2.

In order to simplify the experiments, articles representing the *mixed* class were eliminated so that further research was considered a binary problem. The size of the classes was also reduced to contain 10,000 objects – 5, 340 reliable and 4, 660 unreliable articles – which resulted in a set balanced as the original one.

4 Experimental Evaluation

4.1 Experiment 1 – Max Features Threshold

The first experiment compared two strategies for extracting attributes from texts – regular n-grams and Hollow n-grams. In both cases, the final assignment of weights to individual phrases was carried out using the TF- IDF function and the classification – using MLP.

Figure 1 presents the balanced accuracy achieved by both classifiers for specific values of max features and accuracy obtained by integrating the classifiers into an

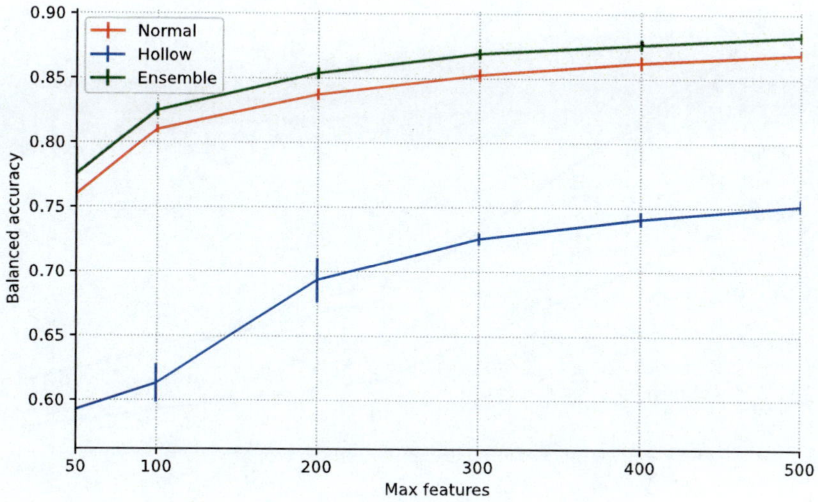

Fig. 1 Accuracy score for all preprocessing strategies and max features values

ensemble with support accumulation. Max features in most cases is an equivalent of representation width – it can be lower only if total number of extracted features is lower. The standard deviations of all test cases supplement results.

It can be observed that bigrams built using the proposed method achieve results from 10 to 15% lower than bigrams using the standard strategy over the entire output space. However, by the assumptions and purpose of the method, the heterogeneous ensemble is characterized by higher accuracy than both component methods, which supports the hypothesis that the use of *Hollow n-grams* allows the extraction of additional information for building models.

4.2 Experiment 2 – Different n values

The second experiment compared the performance of the proposed method, basic approach, and ensemble for different values of n-grams. Not only bigrams were considered, but also strings of three and four words. The maximum number of features after extraction was limited to 500 due to the results achieved in the first experiment.

The study results are presented in Fig. 2, where the noticeable decrease in quality with the increase in the degree of n-grams is visible. It confirms the predictions of Bharadwaj and Shao [2], who speculated that the construction of strings longer than two words will be associated with a decrease in quality due to insufficiently repeatable connections between words and, thus, too little potential for generalization of models.

For the value of $n = 3$, it is also seen that the classifier ensemble achieves worse accuracy than using only basic processing. However, this difference is lower than

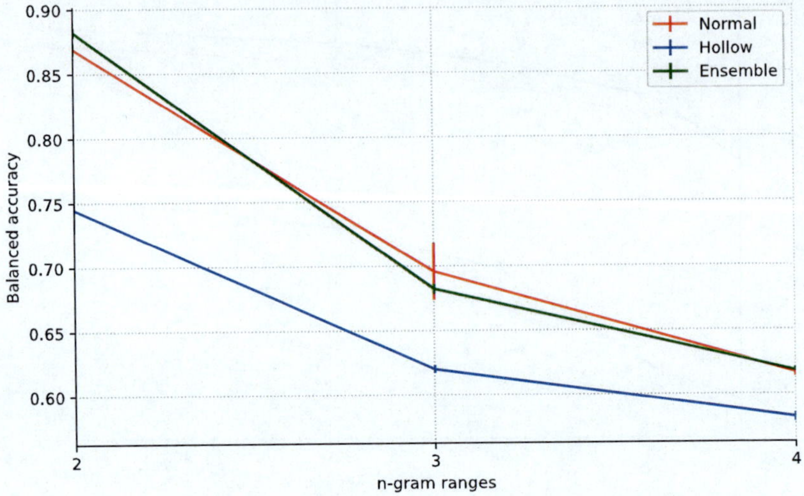

Fig. 2 Accuracy score for all preprocessing strategies and n-gram ranges

standard deviation. In addition, in this case, the proposed method is characterized by a much smaller margin of error than regular n-grams.

As n increases, the difference between the accuracy achieved by Hollow n-grams and the other approaches narrows. In the beginning, it is about 15%, while in the end, the difference is closer to 5%. A slower decline in quality may have its source in omitting the link to the closest word, which involves building the attribute based on further and less specific links.

From the observations, the potential for integrating the indicated preprocessing methods in classifiers ensemble is limited, as is the usefulness of building more complex n-grams. However, *Hollow n-grams* are still potential supplementary information for NLP models, thanks to what it can be used both as a stand-alone method of engineering attributes and supporting the preprocessing procedure for building deep embeds.

5 Conclusions and Future Works

This paper proposed the *Hollow n-grams* method for n-grams building. We assumed it would allow strengthening the generalization abilities of NLP models trained on representations extracted with *state-of-the-art* methods. This hypothesis was substantiated by conducting two experiments, which showed that *Hollow n-grams* has the potential to enhance classification quality for bigram-based ensembles. However, for larger values of n, its usefulness is limited.

In future work, the quality of the classifier ensemble created using the *Hollow n-grams* extraction should be compared with ensemble built from n-grams of several different degrees (e.g., unigrams and bigrams combined) to see if they carry similar generalizing potential.

In addition, much information about the observed phenomena may result from analyzing the source of the potential of the described method to form a classifier ensemble with regular methods. It would allow a better understanding of the mechanisms behind the generalization capabilities of simple NLP methods, which could improve the proposed method.

Acknowledgement This work was supported by the statutory funds of the Department of Systems and Computer Networks, Faculty of Information and Communication Technology, Wroclaw University of Science and Technology.

References

1. Baeza-Yates, R., Ribeiro-Neto, B., et al.: Modern information retrieval, vol. 463. ACM press, New York (1999)
2. Bharadwaj, P., Shao, Z.: Fake news detection with semantic features and text mining. Int. J. Natural Lang. Comput. (IJNLC) **8**, 1–6 (2019)
3. Dale, R.: Gpt-3: what's it good for? Natural Lang. Eng. **27**(1), 113–118 (2021)
4. Fanmuy, G., Fraga, A., Llorens, J.: Requirements verification in the industry. In: Complex Systems Design & Management: Proceedings of the Second International Conference on Complex Systems Design & Management CSDM 2011, pp. 145–160. Springer, Heidelberg (2012). https://doi.org/10.1007/978-3-642-25203-7_10
5. Floridi, L., Chiriatti, M.: Gpt-3: its nature, scope, limits, and consequences. Minds Mach. **30**, 681–694 (2020)
6. Gruppi, M., Horne, B.D., Adalı, S.: Nela-gt-2020: a large multi-labelled news dataset for the study of misinformation in news articles. arXiv preprint arXiv:2102.04567 (2021)
7. Henderson, P., Hu, J., Romoff, J., Brunskill, E., Jurafsky, D., Pineau, J.: Towards the systematic reporting of the energy and carbon footprints of machine learning. J. Mach. Learn. Res. **21**(1), 10039–10081 (2020)
8. Jelodar, H., Wang, Y., Orji, R., Huang, S.: Deep sentiment classification and topic discovery on novel coronavirus or covid-19 online discussions: Nlp using lstm recurrent neural network approach. IEEE J. Biomed. Health Inf. **24**(10), 2733–2742 (2020)
9. Kelleher, J.D., Mac Namee, B., D'arcy, A.: Fundamentals of Machine Learning for Predictive Data Analytics: Algorithms, Worked Examples, and Case Studies. MIT press, Cambridge (2020)
10. Ksieniewicz, P., Choraś, M., Kozik, R., Woźniak, M.: Machine learning methods for fake news classification. In: Intelligent Data Engineering and Automated Learning–IDEAL 2019: 20th International Conference, Manchester, UK, 14–16 November 2019, Proceedings, Part II 20, pp. 332–339. Springer, Heidelberg (2019). https://doi.org/10.1007/978-3-030-33617-2_34
11. Ksieniewicz, P., Zyblewski, P., Borek-Marciniec, W., Kozik, R., Choraś, M., Woźniak, M.: Alphabet flatting as a variant of n-gram feature extraction method in ensemble classification of fake news. Eng. Appl. Artif. Intell. **120**, 105882 (2023)
12. Liu, D., Ye, X.: A matrix factorization based dynamic granularity recommendation with three-way decisions. Knowl. Based Syst. **191**, 105243 (2020)
13. Pedregosa, F., et al.: Scikit-learn: machine learning in python. J. Mach. Learn. Res. **12**, 2825–2830 (2011)

14. Rincy, T.N., Gupta, R.: Ensemble learning techniques and its efficiency in machine learning: a survey. In: 2nd International Conference on Data, Engineering and Applications (IDEA), pp. 1–6. IEEE (2020)
15. Young, I.J.B., Luz, S., Lone, N.: A systematic review of natural language processing for classification tasks in the field of incident reporting and adverse event analysis. Int. J. Med. Inf. **132**, 103971 (2019)

A Non-deep Approach to Classifying Movie Genres Based on Multimodal Data

Paweł Niedziółka and Paweł Zyblewski

Abstract Multimodal data processing has recently become popular due to technological advances and easier access to real video, audio, images, or text data. This data type is often processed using deep neural networks associated with high time and computational complexity. The present work addresses the problem of classifying a multimodal MM-IMDb dataset, representing the problem of recognizing a movie genre based on a poster and a brief description of the plot. For experiments, 20 binary subsets were separated, from which features were then extracted. Features from the text were obtained using the TF-IDF method, while the posters were reduced to a single color. Computer experiments on the resulting tabular data were conducted separately on both modalities, as in the concatenated feature space. The results confirmed that classical approaches to feature extraction could allow satisfactory quality classification of multimodal data to be obtained even when using relatively simple pattern recognition algorithms.

Keywords Data extraction · Multimodal data · Classification

1 Introduction

Data can describe the same object using multiple modalities and can be characterized by images, texts, sounds, or video clips. Modality determines the representation format in which a particular type of information is stored. Multimodal data can provide more information than unimodal data, allowing better analysis and understanding of the information where different sensors are involved [17].

P. Niedziółka (✉) · P. Zyblewski
Department of Systems and Computer Networks, Faculty of Information and Communication Technology, Wrocław University of Science and Technology, Wybrzeże Wyspiańskiego 27, 50-370 Wrocław, Poland
e-mail: 249022@student.pwr.edu.pl

P. Zyblewski
e-mail: pawel.zyblewski@pwr.edu.pl

R. Burduk et al. (eds.), *Progress on Pattern Classification, Image Processing and Communications*, Lecture Notes in Networks and Systems 766,
https://doi.org/10.1007/978-3-031-41630-9_3

Researchers have long studied techniques for combining separate modalities of data into a single feature space. Traditionally, combining data from multiple sensors has been studied from a data fusion perspective; the focus has been on solving the problem of how best to combine data from multiple sources so that the information from different modalities can be preserved [14]. For such a problem, the data extraction process is crucial.

The disadvantage of unsupervised multimodal deep learning methods is that they ignore dependencies between target labels and multiple feature sets when learning multimodal representations [7]. This results in less discriminative representations. Supervised approaches should be used to obtain more discriminative representations because they can use label information when learning representations.

Multi-label data sets are commonly encountered in real-world applications where each instance can simultaneously be associated with multiple labels. To handle such complex data, they are often transformed into binary classification problems where each label is treated separately. This approach allows for using standard binary classifiers, and the decision for each label can be made independently [15]. Additionally, the transformation to binary classification problems can simplify the training process and improve the computational efficiency of the classification algorithm [21].

The paper investigates how classical and simple feature extraction methods for multimodal data perform instead of deep learning, which has a relatively high computational cost. Investigating the classification results for the two modalities separately and combined will answer this.

2 Related Works

This work is related to three popular research directions related to artificial intelligence, namely Natural Language Processing, digital signal/image classification and multimodal data processing.

One of the most popular approaches to text feature extraction is Term Frequency (TF), which measures the frequency of a term in a document [12], while Term Frequency - Inverse Document Frequency (TF- IDF) considers the importance of a term in a collection of documents [2]. Count Vectorizer converts a collection of text documents to a matrix of token counts, which can be used as input to machine learning models [8]. Bag of words is a technique where the frequency of each word in a document is counted and represented as a vector - it is one of the most popular representation methods for object categorization [22]. n-grams are groups of n consecutive words in a sentence and can be used to capture the context of a text. This approach is very accurate in language classification [6]. Natural language processing can also be done using deep learning. The following can be used for this: Convolutional Neural Networks (CNN) that allow the analysis of different features in the data, Recurrent Neural Networks that divide weights differently, unlike the first, and a Long Short-Term Memory Network with mechanisms for retaining, forgetting, and revealing information. Transformers are also increasingly used in encoder-decoder approaches like BERT or GPT [13].

The basic approach to image classification is to extract features using dimensionality reduction methods, such as Principal Component Analysis (PCA), which can also be used for data compression, or data visualization; [10, 18, 19]. Feature descriptors are another way to extract features without using deep learning. They are divided into gradient-based methods such as Scale-Invariant Feature Transform (SIFT), Histogram of Oriented Gradients (HOG), or Dense Daisy feature description (DAISY); intensity-based, Partial Intensity Invariant Feature Descriptor (PIIFD), Local Binary Pattern (LBP), and Binary Robust Independent Elementary Features (BRIEF); or frequency-based spatial methods like Fast Fourier Transform (FFT) [11]. Also known are methods based on point-based feature extraction, such as Centre Surround Extrema Features (CENSURE), which can run in real-time [1]. Another image classification approach is employing CNNs, such as ResNet. Transfer learning is often used in these tasks, where a pre-trained model is fine-tuned on a new data set to improve its performance on a specific task. The tuning involves modifying the weights of the pre-trained model using the new dataset while the early layers remain constant. This approach allows the model to take the knowledge gained during initial training on a large dataset and adapt it to the new task [4].

In machine learning applications, multimodal data includes different modalities, such as video, text, photo, and audio. The authors highlight the potential of combining these modalities to improve the performance of machine learning models [4]. Two main approaches to combining these modalities are early fusion and late fusion. Early fusion involves combining the modalities at the input level before processing them through a model. In contrast, late fusion involves processing the modalities separately through different models and then combining their outputs [5, 20]. The challenges associated with multimodal data are modality alignment and data heterogeneity.

3 Data Preprocessing and Extraction

The dataset used in this study is MM-IMDb [3]. The data was scraped from a popular website that the community used to rate movies, series, or actors, and contains over 50,000 records with two modalities. To begin with, the photos are files in the JPEG format, and they have varying dimensions. While some of these photos are grayscale, others are in color. Secondly, JSON files store text properties like ratings, votes, production companies, plot summaries, and genres. The database presents a multiclass and multi-label problem.

For the purpose of the experiments, the data were extracted and separated into twenty binary imbalanced problems. In the case where a film belonged to both genres, the record was skipped to avoid the same record being classified for two genres. The results are presented in Table 1.

The text data used for the classification task were plots and genres of the movies. The classical extraction was used - the data was tokenized, stopwords were removed,

(a) Action and Sci-Fi

(b) Action and Sci-Fi

(c) Animation and Musical

(d) Animation and Musical

Fig. 1 Example visualization of separated binary datasets in a form o posters and wordclouds (a and c), as well as mean color for each poster in the subset (b and d).

Table 1 Separated binary data subsets

$Class_1$	$\#Class_1$	$Class_2$	$\#Class_2$	IR	Abbr.
Action	2444	Adventure	1604	1.52	ACT- ADV
Action	2768	Sci-Fi	1209	2.29	ACT- SF
Animation	866	Musical	710	1.22	AMT- ML
Biography	1194	Crime	3689	3.09	BIO- CR
Biography	1003	History	803	1.25	BIO- HST
Comedy	6052	Romance	2824	2.14	CD- RC
Documentary	1975	Sport	527	3.75	DOC- SP
Drama	997	Animation	13967	14.01	DRM- AN
Drama	11245	Thriller	2470	4.55	DRM- THL
Drama	13506	Family	1207	11.19	DRM- FML
Family	1668	History	1143	1.46	FML- HST
Family	1608	Mystery	1997	1.24	FML- MST
Fantasy	1881	Short	419	4.49	FNT- SHRT
Horror	2055	Documentary	2676	1.30	HRR- DOC
Mystery	2055	Sport	632	3.25	MST- SP
Romance	1574	Fantasy	5005	3.18	RC- FNT
Sci-Fi	1624	Fantasy	1682	1.04	SF- FNT
Sci-Fi	1946	Short	426	4.57	SF- SHRT
Western	338	Film-Noir	705	2.09	WST- FNR
Western	1093	History	655	1.67	WST- HST

and then the data was vectorized using TF-IDF Vectorizer with the limitation of a maximal number of features to 100. The raw data visualized in Fig. 1a, 1c.

The images were pre-processed to be classifiable; the photos were appropriately downsized (reduction of dimensions) so that they all be the same, and a data extraction operation was performed for the colors - averaging was applied to each of the image's color channels, obtaining, for each poster, a single RGB value. A photo visualization, Fig. 1b, 1d, represents movies posters from two specific genres with a size reduction to RGB pixel. It is worth noting that the visualizations of the posters averaged to a single RGB value indicate the presence of concept drift in the datasets (lighter top rows of the subfigures b and d). This is an important observation when trying to treat separated subsets as data streams. In addition, these datasets may be characterized by a dynamic imbalance ratio [9].

4 Experimental Evaluation

This section provides details of the experiments conducted and the experimental protocol used.

The experiments were designed to answer the following research questions:

1. Which base model achieves statistically significantly best classification performance?
2. Can the fusion of modalities into single feature space result in improved classification accuracy?
3. What is the influence of data preprocessing on the classification accuracy of base models?

4.1 Set-Up

All experiments were conducted for each dataset separately for five different classifiers called with the following parameters (if applicable): Multi-layer Perceptron (MLP) (max number of iteration: 300), K-Nearest Neighbors (KNN) (number of neighbors: 3), Decision Tree (DTC), C-Support Vector (SVC), Gaussian Naive Bayes (GNB). All other parameters default as in sklearn library.

All experiments were performed according to the 5 times repeated stratified 2-fold cross-validation protocol. Results were confirmed using the corrected *student's t test* and Wilcoxon signed-ranks test with $\alpha = 0.05$ [16]. To offset the data imbalance, the Random Over-Sapmling was performed and performance was evaluated using *balanced accuracy score* (BAC). The implementation of the experimental code is publicly available in the GitHub repository[1].

4.2 Experiment 1 – Classification Comparison for Each Modality Separately

Firstly, experiments were conducted for each modality separately - the same generated groups were taken (presented in Table 1), but classifications were made for each modality separately. The results are presented in Table 2.

As can be seen in some cases, the average classification accuracy varies significantly. The indices under the balanced accuracy in Table 2 indicate statistically significantly better classification quality by the classifier than those listed.

If one compares the classification quality for different modalities, it turns out that for this dataset, the classification of textual data performs significantly better than photos.

[1] GitHub repository - https://github.com/KoEj/Multimodal_classifiers.

Table 2 Performance scores of MLP, KNN, DTC, SVC, and GNB

	Photos					Text				
Dataset	MLP[1]	KNN[2]	DTC[3]	SVC[4]	GNB[5]	MLP[1]	KNN[2]	DTC[3]	SVC[4]	GNB[5]
ACT- ADV	0.604 2 3	0.557 –	0.547 –	0.621 2 3	0.614 2 3	0.668 2 3	0.601 –	0.609 –	0.686 2 3	0.682 2 3
ACT- SF	0.543 –	0.535 –	0.523 –	0.558 –	0.539 –	0.696 2 3	0.629 –	0.644 –	0.714 2 3	0.723 1 2 3
AMT- ML	0.516 –	0.548 –	0.550 –	0.509 –	0.503 –	0.678 2	0.618 –	0.644 –	0.685 2	0.681 2 3
BIO- CR	0.603 –	0.574 –	0.557 –	0.635 2 3	0.639 2 3	0.890 2 3 5	0.784 –	0.841 2	0.893 2 3 5	0.868 2 3
BIO- HST	0.606 –	0.535 –	0.534 –	0.633 2 3	0.633 2 3	0.818 2 3	0.726 –	0.743 –	0.810 2 3	0.812 2 3
CD- RC	0.571 –	0.547 –	0.533 –	0.612 2 3	0.601 2 3	0.719 2 3	0.648 –	0.665 –	0.741 1 2 3	0.744 1 2 3
DOC- SP	0.528 –	0.537 –	0.526 –	0.566 3	0.569 2 3	0.729 2 3	0.665 –	0.672 –	0.748 2 3	0.737 2 3
DRM- AN	0.571 –	0.565 –	0.557 –	0.582 –	0.588 –	0.760 2 3	0.662 –	0.728 2	0.762 2	0.783 2 3
DRM- THL	0.546 –	0.543 –	0.542 –	0.564 –	0.521 –	0.751 2 3	0.652 –	0.688 –	0.756 2 3	0.730 2
DRM- FML	0.532 –	0.502 –	0.489 –	0.565 2 3	0.578 2 3	0.818 2 3	0.710 –	0.738 –	0.814 2 3	0.800 2 3
FML- HST	0.524 –	0.522 –	0.509 –	0.535 –	0.543 –	0.838 2 3	0.711 –	0.718 –	0.833 2 3	0.822 2 3
FML- MST	0.586 –	0.539 –	0.523 –	0.594 2 3	0.591 3	0.844 2 3	0.749 –	0.762 –	0.845 2 3	0.845 2 3
FNT- SHRT	0.586 –	0.573 –	0.549 –	0.616 2 3 5	0.597 3	0.740 2 3	0.628 –	0.684 2	0.754 2 3	0.748 2 3
HRR- DOC	0.502 –	0.517 –	0.505 –	0.523 –	0.523 –	0.702 2 3	0.609 –	0.609 –	0.701 2 3	0.678 2 3
MST- SP	0.519 –	0.510 –	0.506 –	0.531 3	0.537 2 3	0.814 2 3	0.736 –	0.762 –	0.824 2 3	0.831 2 3
RC- FNT	0.529 –	0.514 –	0.512 –	0.564 2 3	0.559 2 3	0.677 2 3	0.619 –	0.626 –	0.715 1 2 3 5	0.681 2 3
SF- FNT	0.565 –	0.530 –	0.515 –	0.583 2 3	0.582 2 3	0.854 2 3	0.775 –	0.793 –	0.858 2 3	0.858 2 3
SF- SHRT	0.619 3	0.558 –	0.537 –	0.641 2 3	0.641 2 3	0.626 2 3	0.590 –	0.590 –	0.671 1 2 3	0.695 1 2 3 4
WST- FNR	0.598 2 3	0.532 –	0.517 –	0.602 2 3 5	0.586 2 3	0.614 –	0.579 –	0.575 –	0.627 2 3	0.714 1 2 3 4
WST- HST	0.658 3	0.622 –	0.596 –	0.684 2 3 5	0.673 2 3	0.801 2 3	0.695 –	0.728 2	0.811 2 3	0.801 2 3
Average BAC	0.565	0.543	0.531	0.586	0.581	0.752	0.669	0.691	0.762	0.762
Average rank	3	2.25	1.3	4.45	4	3.5	1.05	1.95	4.4	4.1
	2 3	3	–	1 2 3	1 2 3	2 3	–	2	1 2 3	1 2 3

Table 3 Classification results for multimodal data

Dataset	MLP[1]	KNN[2]	DTC[3]	SVC[4]	GNB[5]
ACT- ADV	0.697 2 3 4	0.557 –	0.618 2	0.619 2	0.688 2 3 4
ACT- SF	0.724 2 3 4	0.535 –	0.641 2 4	0.558 –	0.724 2 3 4
AMT- ML	0.696 2 4	0.548 –	0.630 2 4	0.506 –	0.697 2 3 4
BIO- CR	0.900 2 3 4 5	0.576 –	0.832 2 4	0.633 2	0.873 2 3 4
BIO- HST	0.822 2 3 4	0.536 –	0.725 2 4	0.636 2	0.821 2 3 4
CD- RC	0.749 2 3 4	0.549 –	0.658 2 4	0.612 2	0.750 2 3 4
DOC- SP	0.747 2 3 4	0.537 –	0.676 2 4	0.568 2	0.741 2 3 4
DRM- AN	0.798 2 3 4	0.566 –	0.707 2 4	0.587 –	0.795 2 3 4
DRM- THL	0.746 2 3 4	0.542 –	0.674 2 4	0.560 –	0.746 2 3 4
DRM- FML	0.802 2 3 4	0.504 –	0.734 2 4	0.574 2	0.803 2 3 4
FML- HST	0.837 2 3 4	0.523 –	0.719 2 4	0.545 –	0.839 2 3 4
FML- MST	0.842 2 3 4	0.540 –	0.757 2 4	0.599 2	0.848 2 3 4
FNT- SHRT	0.744 2 3 4	0.573 –	0.678 2 4	0.614 –	0.750 2 3 4
HRR- DOC	0.598 –	0.517 –	0.611 2 4	0.523 –	0.676 2 3 4
MST- SP	0.828 2 3 4	0.512 –	0.745 2 4	0.535 2	0.837 2 3 4
RC- FNT	0.715 2 3 4 5	0.514 –	0.624 2 4	0.564 2	0.683 2 3 4
SF- FNT	0.861 2 3 4	0.532 –	0.794 2 4	0.584 2	0.861 2 3 4
SF- SHRT	0.702 2 3 4	0.559 –	0.598 2	0.642 2 3	0.709 2 3 4
WST- FNR	0.673 2 3 4	0.532 –	0.575 2	0.602 2	0.718 1 2 3 4
WST- HST	0.830 2 3 4 5	0.622 –	0.740 2 4	0.682 2	0.808 2 3 4
Average BAC	0.766	0.544	0.687	0.587	0.768
Average rank	4.35 2 3 4	1.05 –	2.9 2 4	2.1 2	4.6 2 3 4

Table 4 Scores *k*NN and SVC with normalization. STD - StandardScaler, MM - MinMaxScaler, L2 - Normalizer

		NONE[1]	STD[2]	MM[3]	L2[4]
KNN	avg BAC	0.544	0.655	0.650	0.537
	avg rank	1.6	3.8	3.2	1.4
		–	1 3 4	1 4	–
SVC	avg BAC	0.587	0.763	0.778	0.539
	avg rank	1.9	3.075	3.925	1.1
		4	1 4	1 2 4	–

4.3 Experiment 2 – Classification Comparison for Multi-modal Data

Averaged classification accuracy results for each dataset for different types of classification are presented in the Table 3. As in the previous experiment, statistical tests were conducted using the same methods and parameters.

From the data in the table, it can be seen that for the data used in the experiment, the MLP and GNB proved to be the best classifiers. Thus, these are the best methods for classifying the datasets under study. We can also see that combining the available modalities into a single feature space – even despite the low dimensionality of the space resulting from feature extraction from posters – leads to deterioration of KNN and SVC to the level of results achieved on image modality alone. This is likely due to the lack of data preprocessing, which is necessary when using distance-based classification algorithms. This relationship will be investigated in a subsequent experiment.

4.4 Experiment 3 – Impact of Preprocessing on Distance-Based Classifiers Performance

Because of the observations made in Experiment 2, it was decided to repeat the study for distance-based classifiers. This time, however, it was decided to use popular methods of data normalization and standardization, namely L2 normalization (L2), standard normalization (STD), and normalization to [0, 1] range (MM). As can be observed in Table 4, the use of STD and MM allowed for the normalization of the importance of features from both modalities in the concatenated space, and thus, a significant improvement of classification performance. Employing L2 normalization did not achieve this effect, due to the preservation of the weight of features from the image modality in each instance.

5 Conclusion

The work aimed to investigate whether classical approaches to extraction allow for obtaining satisfactory classification quality for a real multimodal problem. The conducted experiments, supported by statistical analysis, proved the validity of using classical approaches to extraction for image and text modalities. In addition, data normalization for some of the classifiers can significantly benefit classification quality.

Future research directions include comparing the quality and processing time with deep methods to verify their necessity. It is also possible to conduct research on multi-label versions of datasets and analyze the problem by treating it as a data stream with concept drift and dynamic imbalance.

Acknowledgments This work was supported by the statutory funds of the Department of Systems and Computer Networks, Faculty of Information and Communication Technology, Wroclaw University of Science and Technology.

References

1. Agrawal, M., Konolige, K., Blas, M.R.: CenSurE: center surround extremas for realtime feature detection and matching. In: Forsyth, D., Torr, P., Zisserman, A. (eds.) ECCV 2008. LNCS, vol. 5305, pp. 102–115. Springer, Heidelberg (2008). https://doi.org/10.1007/978-3-540-88693-8_8
2. Aizawa, A.: An information-theoretic perspective of tf-idf measures. Inf. Process. Manag. **39**(1), 45–65 (2003)
3. Arevalo, J., Solorio, T., y Gómez, M.M., González, F.A.: Gated multimodal units for information fusion (2017)
4. Bayoudh, K., Knani, R., Hamdaoui, F., Mtibaa, A.: A survey on deep multimodal learning for computer vision: advances, trends, applications, and datasets. Visual Comput., 1–32 (2021)
5. Boulahia, S.Y., Amamra, A., Madi, M.R., Daikh, S.: Early, intermediate and late fusion strategies for robust deep learning-based multimodal action recognition. Mach. Vision Appl. **32**(6), 121 (2021)
6. Cavnar, W.B., Trenkle, J.M., et al.: N-gram-based text categorization. In: Proceedings of SDAIR-94, 3rd Annual Symposium on Document Analysis and Information Retrieval, Las Vegas, NV, vol. 161175 (1994)
7. Chen, T., Wang, S., Chen, S.: Deep multimodal network for multi-label classification (2017)
8. Eshan, S.C., Hasan, M.S.: An application of machine learning to detect abusive bengali text. In: 2017 20th International Conference of Computer and Information Technology (ICCIT), pp. 1–6 (2017)
9. Komorniczak, J., Zyblewski, P., Ksieniewicz, P.: Prior probability estimation in dynamically imbalanced data streams. In: 2021 International Joint Conference on Neural Networks (IJCNN), pp. 1–7. IEEE (2021)
10. Kurita, T.: Principal Component Analysis (PCA), pp. 1–4. Springer, Cham (2019). https://doi.org/10.1007/978-3-030-03243-2_649-1
11. Leng, C., Zhang, H., et al.: Local feature descriptor for image matching: a survey. IEEE Access **7**, 6424–6434 (2019)
12. Luhn, H.P.: A statistical approach to mechanized encoding and searching of literary information. IBM J. Res. Dev. **1**(4), 309–317 (1957)

13. Otter, D.W., Medina, J.R., Kalita, J.K.: A survey of the usages of deep learning for natural language processing (2021)
14. Ramachandram, D., Taylor, G.W.: Deep multimodal learning (2017)
15. Read, J., Perez-Cruz, F.: Deep learning for multi-label classification (2014)
16. Stapor, K., Ksieniewicz, P., García, S., Woźniak, M.: How to design the fair experimental classifier evaluation. Appl. Soft Comput. **104**, 107219 (2021)
17. Summaira, J., Li, X., Shoib, A.M., Li, S., Abdul, J.: Recent advances and trends in multimodal deep learning: a review (2021)
18. Topolski, M., Topolska, K.: Algorithm for constructing a classifier team using a modified pca (principal component analysis) in the task of diagnosis of acute lymphocytic leukaemia type b-cll. In: Hybrid Artificial Intelligent Systems (2019)
19. Uddin, M.P., Mamun, M.A., Hossain, D.M.A.: Pca-based feature reduction for hyperspectral remote sensing image classification (2020)
20. Yao, W., Moumtzidou, A., et al.: Early and late fusion of multiple modalities in sentinel imagery and social media retrieval (2021)
21. Zhang, M.L., Zhou, Z.H.: A review on multi-label learning algorithms. IEEE Trans. Knowl. Data Eng. **26**(8), 1819–1837 (2014)
22. Zhang, Y., Jin, R., Zhou, Z.H.: Understanding bag-of-words model: a statistical framework. Int. J. Mach. Learn. Cybern. **1**, 43–52 (2010)

Incremental Extreme Learning Machine for Binary Data Stream Classification

Karol Wojtachnia, Joanna Komorniczak, and Paweł Ksieniewicz

Abstract Classifier ensembles have shown the ability to classify drifted data streams. The following paper proposes an ensemble consisting of a single hidden layer feedforward neural network and an Extreme Learning Machine. For this purpose, a new incremental version of the Extreme Learning Machine is also proposed. Motivations behind such an approach have been precisely described and supported by conducted research. The achieved results show when the architecture might be most useful and what are the possible directions for future development of this method.

Keywords data streams · classification · Extreme Learning Machine · classifier ensemble

1 Introduction

Before choosing or creating a model for machine learning tasks, the critical question is: "What kind of data will it process?". The answer should justify the chosen approach and is essential to get a model that will perform well. In this paper, data streams are the considered type of data.

Based on the literature, the following definition can be formulated: A data stream is a potentially infinite sequence of data items provided in time, with no control over the order in which they arrive [1]. Storing the whole stream in memory is impossible, and the processing systems should be capable of giving a response at any time [5]. Processing of streams can be done sample by sample (online processing), or models are trained iteratively based on subsets of the data stream (batch processing). Each batch contains several most recent examples [7]. Data streams are to be found in many essential areas of the industry: analysis of computer networks, monitoring logs of cybersecurity systems, or processing responses from sensors.

K. Wojtachnia (✉) · J. Komorniczak · P. Ksieniewicz
Wrocław University of Science and Technology, Wrocław, Poland
e-mail: karol.wojtachnia@pwr.edu.pl

R. Burduk et al. (eds.), *Progress on Pattern Classification, Image Processing and Communications*, Lecture Notes in Networks and Systems 766,
https://doi.org/10.1007/978-3-031-41630-9_4

Data Streams. The described data type creates new constraints and challenges for pattern recognition tasks. First, a potential change in the posterior data distribution should be considered. Concept drifts can be related to class distribution change, decision boundary shifts, or prior probability transitions in the incoming data.

From the perspective of the time over which drifts occur, they can be divided into abrupt, gradual, and incremental [16]. Sudden drift occurs when a concept change occurs abruptly from one chunk to another. There is no intermediate state between the initial and final concepts. A single moment in the stream can be identified, after which all objects belong to the new concept. Gradual drift occurs over time and manifests itself as a change at a slow pace, with changing probability of observing samples from two consecutive concepts. In other words, a period can be distinguished in which some samples in chunk represent an old concept and others a new concept. As time passes, there are more and more chunks with the new concept until, finally, the old concept disappears completely. Incremental drift behaves similarly, but the change occurs smoothly – the transition from the old to the new concept is continuous. The objects in a given chunk come from a temporary transitional concept during the transition. However, over time the share of chunks from the new concept increases until the initial one disappears completely. It should also be noted that individual deviations from a concept do not describe concept drifts. Hence, noise and outliers should not be counted as such.

A drift will be recurrent when an old concept reappears in the stream after some time. If, on the other hand, a change in the distribution of examples in the decision space does not cause a change in the decision boundary, then we are dealing with a virtual drift [11]. A separate difficulty is the possibility of the emergence of a new class and drifts of prior probability over the stream course [10].

Concept Drift Adaptation. There are three approaches to dealing with concept drift. The first assumes frequent rebuilding of the entire model. Due to the computational complexity, this way may be impractical and inefficient. Alternatively, one can use drift detection and retrain the model if the changes are significant enough. The method proposed in this work falls into the third category. The model is continuously trained to adapt to the changes occurring in the decision-making environment [12].

A required feature of methods dedicated to data stream processing is constant memory complexity. It can be achieved when the model can process each sample only once, which makes storing them supererogatory [5]. Furthermore, training and prediction times should remain limited.

This paper proposes a modification of an Extreme Learning Machine (ELM) suitable for data stream processing. The method meets all the requirements stated above.

2 Methods

Extreme Learning Machine (ELM) was proposed as an alternative for gradient-based learning where all the network parameters are tuned iteratively [9]. It is a feedforward neural network with only one hidden layer. The method's learning process starts with setting random values to all weights connecting the input and hidden layer as well as biases in the hidden layer. The end weights between the hidden and output layers are calculated using the Moore-Penrose inverse transform.

Equation 1 describes the last step of the learning process.

$$\beta = (\frac{1}{C} + H^T H)^{-1} H^T T \tag{1}$$

C is a diagonal matrix adding the same positive value c (which is a parameter of the network) in order to achieve better generalization [15]. H contains values of neurons in the hidden layer calculated for every sample in the training set, and T contains expected output for corresponding training samples.

Using an *Extreme Learning Machine* in data stream classification could potentially master new concepts much faster than a traditional *Single hidden layer feedforward neural network* (SLFN). It is also expected that an SLFN will achieve better classification *accuracy* than incremental ELM, some chunks after the drifts occur. Based on these hypotheses, an iterative learning process has been proposed, and a simple ensemble containing ELMi (as the proposed incremental version of ELM is called in this paper) and SLFN to combine their advantages and minimize the disadvantages.

The proposed iterative learning of ELM goes as follows. With the first chunk, the model is taught like a classical ELM. For every new chunk, calculate another set of weights β_n based only on new examples. Then, with some factor Δ, calculate weighted average of current β and β_n. Set the calculated average as β values of the model. This process can be called inertial smoothing of coefficient β for streaming data classification tasks.

In recent times, it has become increasingly popular to use classifier ensembles instead of just one. The rationale behind this is that using a more diverse set of classifiers makes it possible to achieve a higher degree of *accuracy* in classification [14]. A classifier ensemble, also known as Multiple Classifier System [4], comprises two essential components: a set of cooperating classifiers and a combination rule that generates the final prediction based on individual decisions made by each model in the ensemble [2]. The proposed ensemble employs a combination rule that is based on the *accuracy* metric. After each evaluation step, the classifier with the highest score is identified. Until the next evaluation occurs, the ensemble will always return the prediction made by the model that performed better in the last evaluation step. The process is also described by Algorithm 1.

Algorithm 1: Pseudocode of the learning process with proposed ensemble

$\mathcal{DS} = \{\mathcal{DS}_1, \mathcal{DS}_2, \ldots, \mathcal{DS}_k\}$ – data stream,
Φ_M – Multilayer Perceptron with one hidden layer
Φ_E – incremental Extreme Learning Machine
ρ – logic variable holding information whether prediction is done by Φ_M or Φ_E
$\rho \leftarrow false$ **forall the** $\mathcal{DS}_k \in \mathcal{DS}$ **do**
end
if $\mathcal{DS}_k$ *is not a first data chunk* **then**
end
Evaluate both models
$A_M \leftarrow accuracy(\Phi_M, \mathcal{DS}_k)$
$A_E \leftarrow accuracy(\Phi_E, \mathcal{DS}_k)$
Store information which model performed better **if** $A_M > A_E$ **then**
end
$\rho \leftarrow True$ **else**
end
$\rho \leftarrow False$
Classifier fitting
fit Φ_M with $\mathcal{DS}_k$
fit Φ_E with $\mathcal{DS}_k$

3 Evaluation of Data Stream Classifiers

When computational resources are limited, the characteristics of data streams make cross-validation insufficient for evaluating predictive performance [8]. An example of a suitable approach could be holding out an independent test set and applying the recent model to it at regular intervals. This way is unsuitable for drifting streams as it does not consider potential changes in data distribution. Considering the batch analysis method, one approach appropriate for drifted streams is *Test-Then-Train* [16], where the stream is processed sequentially chunk by chunk. On each chunk apart from the first one, the classifier is first evaluated, and then the model is updated based only on the currently processed batch. Another method is a window sliding over the stream and is called *Prequential Analysis* [6]. After evaluating and updating the model, the window moves on, discarding some of the old data and including an equal number of new samples in the collection.

3.1 Experimental Setup

This section describes the procedure conducted to gather data needed for evaluation. Also, the conducted experiments are explained in detail.

The stream generator from the *stream-learn* [13] library was utilized to create synthetic quantitative data streams. This generator allows the user to set parameters such as the number of drifts, the type of drifts, and the number of chunks. In every experiment, binary problems described by 250 chunks, each containing 1000 objects, were employed. Each object was described by 20 informative features. In the tests, neural networks contained 100 neurons in the hidden layer.

To begin with, the performance of incremental ELMs with different values of learning factor was examined. For this purpose, the following values of the delta were used: 0.3, 0.5, 0.8. At first, these classifiers were tested on streams that included nine gradual drifts. Then similar tests were conducted with 18 sudden drifts. *Accuracy* metric has been measured on each chunk.

Secondly, a comparison was made between ELMi and MLPs with the same number of neurons in the hidden layer to determine the difference in classification accuracy and restoration curve. The neural network implementation from the scikit-learn library [3] was utilized for this assessment. To gather needed data, streams with 9, 18, and 24 drifts were created for two types of drift: gradual and sudden.

The stream generator used in experimental evaluation operates on drifts occurring in a fixed interval. This allows for internal replication of the model's performance in relation to drift by pleating the vector of quality flow and averaging along the fold axis.

4 Experimental Evaluation

This part of the work concentrates on presenting the results obtained in the experiments as graphs and tables. The observations made based on these make it possible to assess the quality of the proposed models and to check the validity of their application.

4.1 ELMi Hyperparameters

As a first step, the experiment aimed to investigate the behaviour of the incremental ELM for different values of the Δ coefficient. The left-hand graph in Fig. 1 shows the averaged results of the *accuracy* metric over individual chunks in the gradual drift stream. It can be seen that after each drift occurrence, the classifier learns a new concept and stabilizes its results after only a few chunks. It should also be noted that after temporary drops in quality due to drifts, the results return to the interval (0.625, 0.650) each time. Analogical results were obtained for the same tests but performed on a stream with an abrupt drift, as shown in the second part of the graph of Fig. 1. None of the graphs shows significant differences in the results obtained by models with different learning rates. It is worth observing that, in both cases, the dynamics of decrease and increase in classification quality are analogous to the dynamics of the emergence of new concepts. For abrupt drifts, sudden decreases and very rapid increases in quality were obtained, while for gradual drifts these are more smooth.

The plot of Fig. 2 shows the averaged quality curves for the chunks in which model reconstruction occurs after a drift. Although, on average, the maximum and

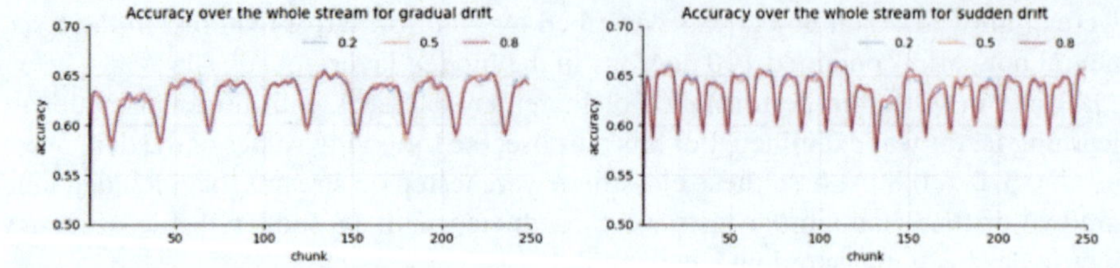

Fig. 1 Comparison of *accuracy* by ELMi with different update rates for gradual (left) and sudden (right) drift

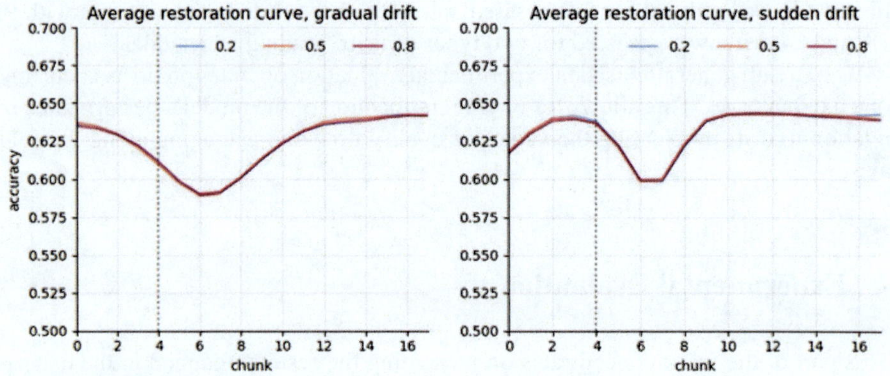

Fig. 2 Comparison of average restoration curve for different update factors in case of gradual drift

minimum quality in both cases are very similar, it can be seen that for abrupt drifts, the models react faster and update their parameters to the new concept quicker. As it was apparent from the study that the smoothing factor (delta) did not affect the results, a value of 0.8 was set in further studies.

4.2 Comparison of Classifiers

The second section focuses on comparing the incremental ELM, MLP, and the ensemble of classifiers created with them. As seen in the left plot of Fig. 3, the MLP maintains quality better than ELMi at all times in the case of less frequent gradual drifts, while the ensemble behaves the same, as it always returns the prediction given by the classically trained network. When looking at the right plot of Fig. 3, one can see moments when, after a drift has occurred, ELMi shows better classification quality than MLP for a certain number of chunks. For this reason, we can see a slightly better accumulated *accuracy* achieved by the ensemble.

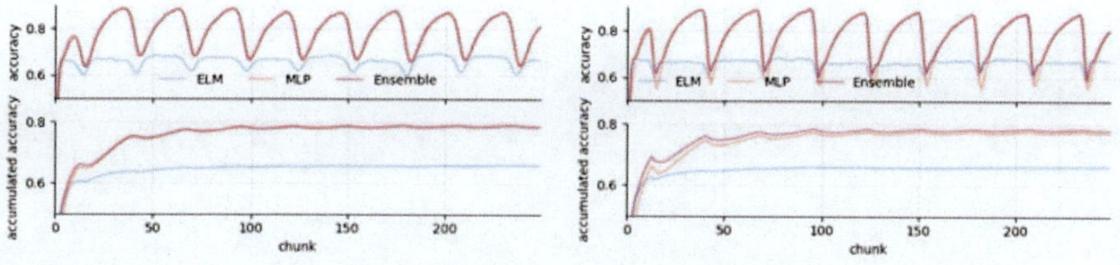

Fig. 3 Comparison of *accuracy* by ELMi, MLP and ensemble in streams with 9 gradual (left) and sudden (right) drifts

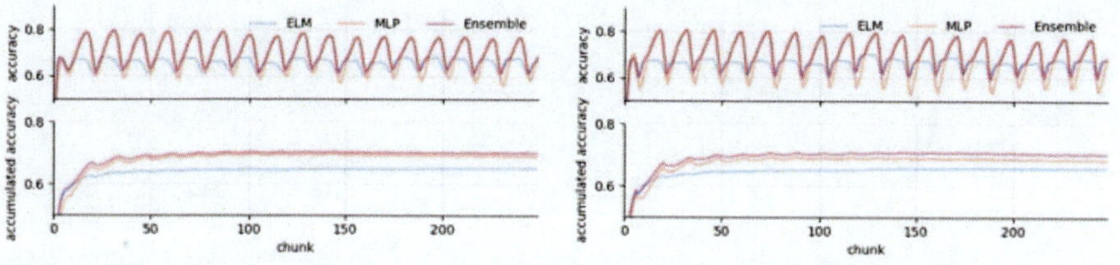

Fig. 4 Comparison of *accuracy* by ELMi, MLP and ensemble in streams with 18 gradual (left) and sudden (right) drifts

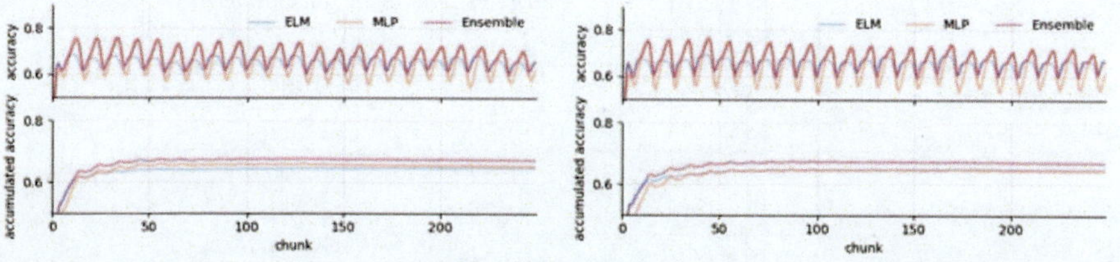

Fig. 5 Comparison of *accuracy* by ELMi, MLP and ensemble in streams with 24 gradual (left) and sudden (right) drifts

By increasing the number of drifts for the same stream length, an apparent decrease in the average classification *accuracy* of the MLP can be observed in Fig. 4. As the average *accuracy* of ELMi remained at a similar level, the apparent decrease in *accuracy* for the ensemble is as expected. The difference between MLP and ensemble performance in the graph is again slightly higher for sudden drift.

Figure 5 illustrates the performance of the three models for very common drifts. Again, a decrease in classification quality can be observed for MLP and ensemble. Interestingly, ELMi still achieves average results around 0.7. It should also be noted that the average accumulated *accuracy* MLP and incremental ELM values equalize for such frequent drifts.

In Fig. 6 average restoration curves have been presented. They clearly show that incremental ELM returns faster to the previous classification quality than MLP with the same architecture. The observations do not apply when nine gradual drifts were

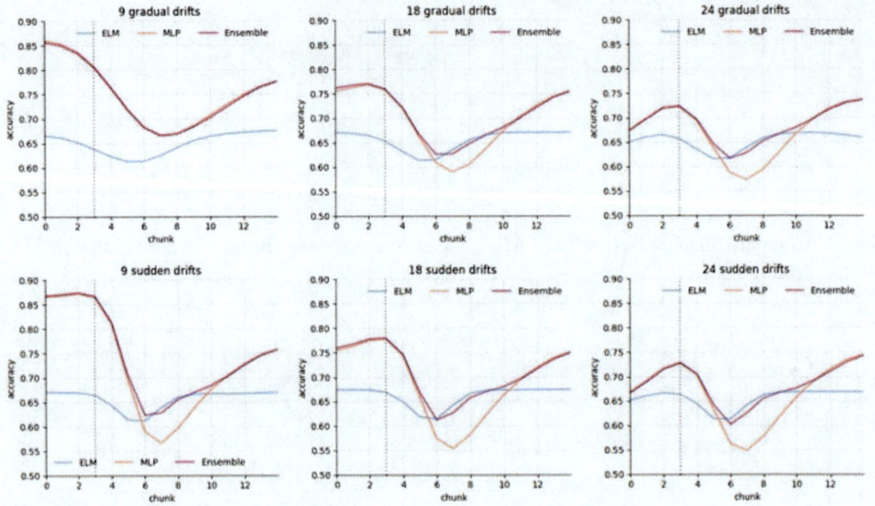

Fig. 6 Comparison of average recovery curve for gradual (top) and sudden (bottom) drifts

Table 1 avg. Accuracy for gradual drift

Drifts	ELMi	MLP	ensemble
	1	2	3
9	0.657	0.785	0.784
	–	1	1
18	0.651	0.689	0.700
	–	1	1,2
24	0.646	0.651	0.671
	–	1	1,2

tested. Only then MLP was able to recover fast enough to outperform ELMi all the time. In every other case, it takes on average about four chunks for MLP to outperform ELMi again after it drops below incremental ELMs performance.

Table 1 and Table 2 present the results of paired Student's t-test for the *accuracy* metric averaged over the entire length of tested streams. In almost all cases MLP and the ensemble have proven to be statistically significantly better than ELMi. For 18 and 24 drifts ensemble was better than MLP regardless of drift type. However, absolute differences are very small as shown in the plots. Especially for 24 sudden drifts, statistical significance exists between results reached by ELMi and MLP but the absolute difference is only 0.007.

Table 2 avg. Accuracy for sudden drift

Drifts	ELMi	MLP	ensemble
	1	2	3
9	0.660	0.769	0.778
	–	1	1
18	0.658	0.683	0.702
	–	1	1,2
24	0.649	0.642	0.669
	2	–	1,2

5 Conclusions

In this work, an incremental modification of the Extreme Learning Machine was proposed and implemented, as well as an ensemble combining it with a traditionally trained MLP. The solution fulfills requirements currently placed on classifiers working with drifted streams. The data collected and the observations have led to the following conclusions:

- For the research carried out so far, the ensemble is an added value, especially for sudden drifts.
- If drifts are rare and gradual, results obtained so far for the ELM and ensemble do not justify their use in such an environment.
- ELMi achieves very similar accumulated *accuracy* regardless of drift type and frequency of its occurrence.
- On the basis of the research carried out, it can be concluded that ELMi does not give outstanding results on its own. However, using the classifier in an ensemble allows it to be used when the MLP has not yet adapted to the new concept.
- Apart from rare (9) gradual drifts, the ensemble achieved the best results. The differences were not large, but this offers the prospect that the results would be even better with a better selection of ELM parameters.

Several perspectives can be explored to further advance the development of ELMi as an ensemble learning method. Firstly, there is a need to test the model for other values of ELM parameters such as C and the number of neurons in the hidden layer. Secondly, investigating the behaviour and comparing the performance of the tested models for even longer streams will provide a deeper understanding of the capabilities and limitations of ELMi. Thirdly, experimental testing of the learning time and prediction time of ELMi is necessary to determine the cost of using the method in an ensemble. Finally, the proposed ensemble combination method relies on having a labelled set, which may not always be feasible. Therefore, finding another combination method that is less susceptible to this requirement is worth evaluating.

The observations presented in this paper and the obtained results justify further research on the presented methods, which will help determine even more confidently whether the developed solutions have applications in processing drifting data streams.

Acknowledgement This work was supported by the statutory funds of the Department of Systems and Computer Networks, Faculty of Information and Communication Technology, Wroclaw University of Science and Technology.

References

1. Agrahari, S., Singh, A.K.: Concept drift detection in data stream mining?: a literature review. J. King Saud Univ. Comput. Inf. Sci. **34**(10, Part B), 9523–9540 (2022)
2. Britto, A.S., Jr., Sabourin, R., Oliveira, L.E.: Dynamic selection of classifiers-a comprehensive review. Pattern Recogn. **47**(11), 3665–3680 (2014)
3. Buitinck, L., et al.: API design for machine learning software: experiences from the scikit-learn project. In: ECML PKDD Workshop: Languages for Data Mining and Machine Learning, pp. 108–122 (2013)
4. Burduk, R.: Integration base classifiers based on their decision boundary. In: Rutkowski, L., Korytkowski, M., Scherer, R., Tadeusiewicz, R., Zadeh, L.A., Zurada, J.M. (eds.) ICAISC 2017. LNCS (LNAI), vol. 10246, pp. 13–20. Springer, Cham (2017). https://doi.org/10.1007/978-3-319-59060-8_2
5. Domingos, P., Hulten, G.: Mining high-speed data streams. In: Proceedings of the Sixth ACM SIGKDD International Conference on Knowledge Discovery and Data Mining, pp. 71–80 (2000)
6. Gama, J., Sebastião, R., Rodrigues, P.: On evaluating stream learning algorithms. Mach. Learn. **90**, 317–346 (2013)
7. Gama, J., Žliobaitė, I., Bifet, A., Pechenizkiy, M., Bouchachia, A.: A survey on concept drift adaptation. ACM Comput. Surv. (CSUR) **46**(4), 1–37 (2014)
8. Grzyb, J., Klikowski, J., Woźniak, M.: Hellinger distance weighted ensemble for imbalanced data stream classification. J. Comput. Sci. **51**, 101314 (2021)
9. Huang, G.B., Zhu, Q.Y., Siew, C.K.: Extreme learning machine: a new learning scheme of feed-forward neural networks. In: 2004 IEEE International Joint Conference on Neural Networks (IEEE Cat. No. 04CH37541), vol. 2, pp. 985–990. IEEE (2004)
10. Komorniczak, J., Zyblewski, P., Ksieniewicz, P.: Prior probability estimation in dynamically imbalanced data streams. In: 2021 International Joint Conference on Neural Networks (IJCNN), pp. 1–7. IEEE (2021)
11. Krawczyk, B., Minku, L.L., Gama, J., Stefanowski, J., Woźniak, M.: Ensemble learning for data stream analysis: a survey. Inf. Fusion **37**, 132–156 (2017)
12. Ksieniewicz, P., Woźniak, M., Cyganek, B., Kasprzak, A., Walkowiak, K.: Data stream classification using active learned neural networks. Neurocomputing **353**, 74–82 (2019)
13. Ksieniewicz, P., Zyblewski, P.: Stream-learn – open-source python library for difficult data stream batch analysis. Neurocomputing **478**, 11–21 (2022)
14. Kuncheva, L.I.: Combining Pattern Classifiers: Methods and Algorithms. John Wiley & Sons, Hoboken (2014)
15. Zong, W., Huang, G.B., Chen, Y.: Weighted extreme learning machine for imbalance learning. Neurocomputing **101**, 229–242 (2013)
16. Zyblewski, P.: Classifier selection for imbalanced data stream classification (2021)

Interpreting Intrusions - The Role of Explainability in AI-Based Intrusion Detection Systems

Marek Pawlicki, Aleksandra Pawlicka, Mścisław Śrutek, Rafał Kozik, and Michał Choraś

Abstract Machine learning has become a key component of the effective detection of network intrusions. Yet, it comes with the lack of transparency - an issue which can be mitigated with the employment of explainable AI techniques. In this paper, the crucial role of explainability in intrusion detection is discussed, along with its benefits and drawbacks, followed by presenting and comparing the results of four main explainability techniques applied to an intrusion detection system.

Keywords cybersecurity · explainability · explainable AI · intrusion detection · xAI

1 Introduction, Motivation and Rationale

In today's interconnected world, heavily relying on the Internet and related services, and threatened by a wide variety of cyberattacks, intrusion detection systems (IDS), which provide real-time monitoring and are capable of uncovering any suspicious activity, have become an inseparable component of the network security paradigm. Yet, modern networks are becoming progressively more complex, and the sophistication of cyberattacks constantly rises. As a result, it has been increasingly more difficult for IDSs to detect and classify intrusions in an accurate manner [11].

One way to tackle this problem of IDSs has is with the use of machine learning (ML) algorithms; they have shown promising results in improving detection accuracy. Still, these complex algorithms tend to be opaque, i.e., they work in a black-box manner. This means that human operators are unable to understand how a particular

M. Pawlicki (✉) · A. Pawlicka · R. Kozik · M. Choraś
ITTI, Poznań, Poland
e-mail: marek.pawlicki@pbs.edu.pl

M. Pawlicki · M. Śrutek · R. Kozik · M. Choraś
Bydgoszcz University of Science and Technology, Bydgoszcz, Poland

A. Pawlicka
University of Warsaw, Warsaw, Poland

R. Burduk et al. (eds.), *Progress on Pattern Classification, Image Processing and Communications*, Lecture Notes in Networks and Systems 766,
https://doi.org/10.1007/978-3-031-41630-9_5

decision was made, and why [16]. In a situation where it is impossible to achieve transparency of the decision-making process, concerns are raised about the reliability and trustworthiness of the models. This issue becomes all the more important in high-stakes domains, such as healthcare, autonomous vehicles, finances, and, as in the case of this work, cybersecurity [2, 10].

This lack of transparency gave rise to the postulate of making AI explainable; thus, the concept of explainable AI (xAI) was introduced. Explainability refers to the humans' ability to understand and interpret the decision-making process of ML algorithms. In the context of intrusion detection, explainability may help identify the critical features of the data that contribute to a particular decision, making it easier for human analysts to understand the nature of the threat and develop effective countermeasures. On top of that, explainability can lead to identifying biases and errors in the models, resulting in their improved performance [15].

This paper highlights the pivotal role of explainability in intrusion detection systems and discusses the challenges and opportunities associated with the use of xAI techniques in this domain. The most relevant xAI methods are briefly presented. Then, the results of an experiment are shown, in which a number of state-of-the-art explainability techniques are employed in an intrusion detection system. The results are juxtaposed, with the conclusions drawn thereafter.

2 Materials and Methods

xAI techniques are being increasingly used in intrusion detection. These techniques offer insights into the decision-making processes of ML algorithms, facilitating human understanding of the rationale behind a particular decision.

2.1 Advantages and Limitations of xAI Techniques

There are several advantages to intrusion detection that the application of xAI techniques brings about. First of all, the xAI techniques can shed light on the decision-making process of ML models. Thanks to this, it becomes easier for human analysts to understand how and why a particular decision was made. This further leads to improved trust in the ML models; consequently, their adoption in real-world settings could increase as a result [8]. Additionally, xAI techniques can help to identify biases and errors in the ML models, the correction of which will positively influence their performance [3].

It has also been argued that xAI techniques facilitate the development of more effective and efficient countermeasures against potential threats by providing insight into the factors that contribute to a particular decision. xAI techniques may help to find and pinpoint the critical aspects of the data that require attention when investigating potential threats. This can lead to more targeted and precise investigations, ultimately saving time and resources [4].

The application of xAI techniques has the potential to facilitate collaboration between human analysts and ML models. Specifically, xAI techniques provide interpretable representations of the decision-making process, which enables human analysts to validate the decisions made by ML models as well as provide feedback for improvement. This results in a more effective partnership between human analysts and ML models, ultimately improving the overall performance of the intrusion detection system [4].

Yet, xAI techniques are not free of potential shortcomings. One such drawback is the trade-off between interpretability and accuracy. This means that making models interpretable in order to understand the decision-making process may impede their level of accuracy, in comparison with the more complex, black-box models [3].

Despite these drawbacks, xAI techniques have shown promise in improving the transparency and interpretability of ML models in intrusion detection. In the context of intrusion detection, xAI techniques can provide insights into the decision-making process, identify biases and errors, and facilitate collaboration between human analysts and ML models [8].

There are a number of xAI techniques which could be employed in intrusion detection; some of the most prominent ones have been discussed below.

2.2 *Overview of xAI Techniques Applicable in Intrusion Detection*

Model-agnostic techniques provide local explanations for individual predictions. As the name suggests, they can be applied to any ML model. In the context of intrusion detection, the methods of this kind allow security analysts to better understand how a particular prediction was made. This is particularly valuable when the models are complex, making it fairly possible for humans to interpret. Some of the most recognizable techniques of this kind are LIME, SHAP; DiCE and ProtoDash.

LIME (Local Interpretable Model-Agnostic Explanations) is a technique that generates locally faithful explanations for individual predictions by fitting an interpretable model to the neighbourhood of the input data, thus providing an insight into a complex model by generating a much simpler, locally interpretable linear model [12].

SHAP (SHapley Additive exPlanations) is a technique that assigns a value to each input feature based on its contribution to the model's prediction, leveraging cooperative game theory. The method exploits Shapley values, which are used to fairly allocate the payoff among players in a coalition, treating features as players and the model output as payoff [7].

DiCE (Diverse Counterfactual Explanations) provides counterfactual explanations - informative samples which answer the question 'What would it take to change the label on this sample?'. DiCE synthesises those data points by perturbing features of a sample until the label flips, essentially presenting a 'what-if' scenario. The particular way of finding the data points is an optimisation problem similar to the generation of adversarial attacks, perturbing the features which maximise the effect on the label and minimise the amount of necessary perturbation [9]. DiCE Counterfactuals are a valuable supplement to other xAI techniques.

ProtoDash allows to find 'Prototypical Samples' of a set of data, which, for a labelled dataset, can be used to gain insight into the characteristics of a subset, or a specific class of data, through its most representative samples. ProtoDash provides a concise representation of the given class. The samples are found by maximising similarity [5].

3 Experimental Setup and Results

The experiment was carried out on an open, public cybersecurity benchmark, the CICIDS2018 subset of the Netflow Collection, which is a conversion of the CICIDS2018 [14] dataset to the Netflow format [13]. Before using the dataset, the duplicate values were dropped and the categorical features were encoded via Frequency Encoding. The set was split into training and testing set (80/20), stratifying the sets by the target label. The training set was balanced using random subsampling on the majority (Benign) class to the number of the sum of samples of all the attack classes, achieving 1:1 ratio attack:benign. Before inputting to the neural network, the features are scaled using scikit-learn's StandardScaler.

A neural network is trained on the training set, the hyperparameters are listed in Table 1.

Table 1 Hyperparameters of the Neural Network Model

Hyperparameter	Value
Optimizer	Adam
Learning Rate	0.0001
Hidden Layers	2
Units in Layer 1	360
Activation in Layer 1	ReLU
Units in Layer 2	360
Activation in Layer 2	ReLU
Output Activation	Softmax
Loss Function	Categorical Crossentropy
Batch Size	2
Epochs	50
Early Stopping Monitor	Loss
Early Stopping Patience	5

Table 2 Results of the classification model

	Accuracy	BACC	MCC	Precision	Recall	Hamming Loss
Results	0.873	0.441	0.795	0.853	0.873	0.127

The hyperparameters were chosen using the hyperband algorithm [6]. The used framework is TensorFlow [1]. The model's performance is evaluated using a standard set of metrics, contained in Table 2.

In addition to Accuracy, Precision and Recall, Hamming loss, Matthews Correlation Coefficient and Balanced Accuracy are calculated to give a better overview of the model performance. Hamming loss measures the number of mistakenly classified samples to the total number of samples, estimated as the hamming distance between y_true and y_predicted. The measure ranges from 0 to 1, with 0 being the better classifier.

The Matthews Correlation Coefficient (MCC) takes into account all the elements of a confusion matrix: true positive (TP), true negative (TN), false positive (FP), and false negative (FN), providing a balanced assessment of classification. The score ranges from -1 to 1, with 0 being no better than random chance, 1 representing the perfect score, and -1 the opposite. The equation for MCC is given in Eq. (1).

$$MCC = \frac{(TP \times TN) - (FP \times FN)}{\sqrt{(TP + FP)(TP + FN)(TN + FP)(TN + FN)}} \tag{1}$$

Balanced Accuracy is a metric particularly useful in imbalanced datasets, it is the average of sensitivity and specificity, as displayed in Eq. (2).

$$BACC = \frac{1}{2}\left(\frac{TP}{(TP + FN)} + \frac{TN}{(TN + FP)}\right) \tag{2}$$

With the detection model established and evaluated, the experiment proceeded with attempting to gain insight into the decision making process of the classifier. To this end, the four xAI methods were employed: SHAP, LIME, DiCE and ProtoDash. Following a scenario where a security operative needs to justify a decisive action to ban or not to ban a user based on the detection result, the focus will be on local explanations, which will, in principle, provide reasoning as to why the samples in question were classified as an attack.

3.1 xAI Interpretation

An attack was randomly selected from the attack classes; the model predicted the sample to be a 'DOS-Hulk' attack. The analysis begins with the SHAP summary plot, which showcases the feature importance calculated for the entire test set, as seen in

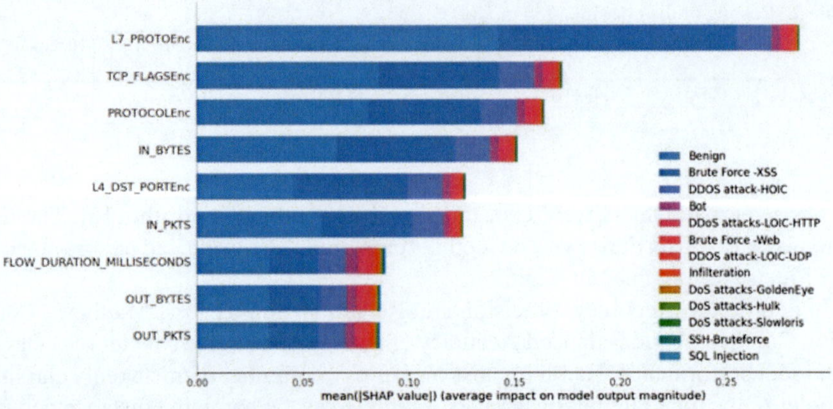

Fig. 1 SHAP: average impact on model output

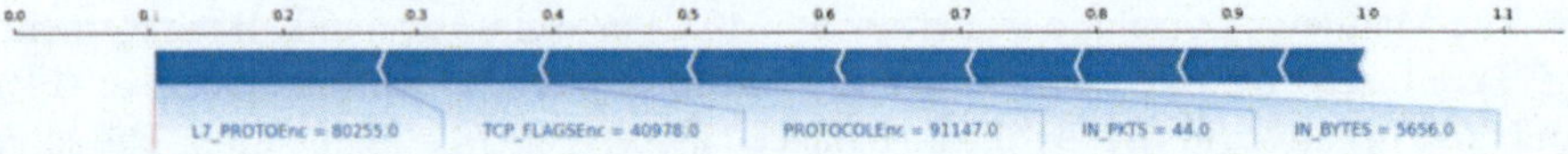

Fig. 2 SHAP: contributions of features to the classification for the specific 'DOS-Hulk' sample

Fig. 1. The feature importance is sorted, and its contribution to particular classes is depicted in colour. With the general overview of the behaviour of the model, the SHAP force plot is used to understand the contributions of features to the particular classification of the 'DOS-Hulk' sample, as seen in Fig. 2. The force plot falls in line with the summary plot, with 'L7 Proto' being the strongest feature, closely followed by 'TCP FLAG'. However, this explanation is contradicted by the one coming from LIME in Fig. 3, where the local interpretable linear model considered 'Flow Duration' and 'Out Packets' to be the two most contributive features. Following LIME, ProtoDash was used to gain insight into the characteristics of the 'DoS-Hulk' class. In Table 3, three prototypical samples selected by the algorithm are showcased, along with their feature values and the weight - the relevance of the sample with respect to the representation of the explained instance class. The explained sample is given at the bottom row of the table. Finally, DiCE is employed to find counterfactual samples, which are listed in Table 4, with the explained sample listed at the bottom of the table. The target label of the DiCE algorithm was selected to be 'Benign', so the found counterfactuals answer the question 'What would have to change so this sample is classified as Benign'. It is noteworthy that of the five proposed counterfactuals, three have a very similar flow duration to the explained sample, which contradicts the immense importance attributed to that value by the LIME algorithm.

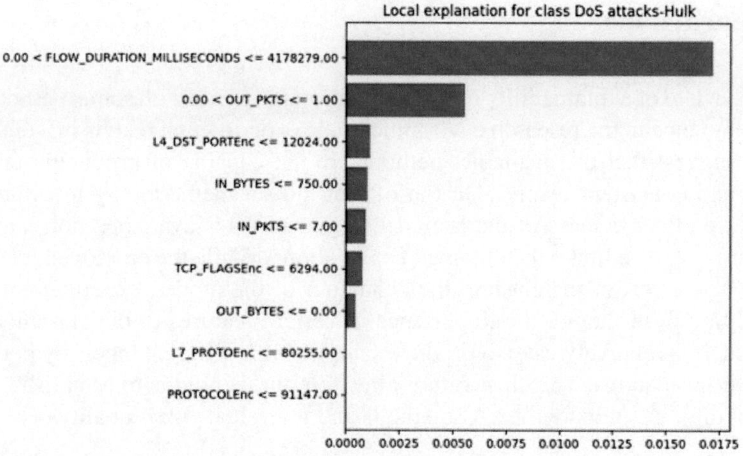

Fig. 3 LIME for a sample from the DOS-Hulk class

Table 3 Protodash Prototypes selected by the algorithm for the sample given at the bottom of the table

	INB	OUTB	INP	OUTP	FLOW DUR.	L7	PROT	DST PORT	TCPFL	Weight	Label
0	172.0	2861.0	4.0	68.0	4294942.0	0	6	3389	219	0.981968	DoS attacks-Hulk
1	743.0	1269.0	6.0	6.0	4294905.0	92	6	22	219	0.009456	DoS attacks-Hulk
2	848.0	2264.0	11.0	10.0	4238765.0	83	6	63991	219	0.007987	DoS attacks-Hulk
	172	2821	4	9	4294949	0	6	3389	219	–	DoS attacks-Hulk

Table 4 Counterfactual Samples synthesized with DiCE

	INB	OUTB	INP	OUTP	FLOW DUR.	L7	PROT	DST PORT	TCPFL	Label
0	14496782.0	2103.0	13.0	9.0	3685515.2	0.0	6	3389	219	Benign
1	4609669.5	2103.0	13.0	9.0	4291385.0	0.0	58	3389	219	Benign
2	1991.0	2103.0	36057.535	9.0	4291385.0	0.0	6	3389	219	Benign
3	1991.0	70182360.0	13.0	9.0	3222583.0	0.0	6	3389	219	Benign
4	1991.0	2103.0	13.0	9.0	4291385.0	0.0	6	3389	4	Benign
Samp.	172	2821	4	9	4294949	0	6	3389	219	DoS attacks-Hulk

4 Conclusions

While the field of explainability of AI is a hot research topic, with some methods gaining prominence in the research environment, the experimental results in a real-world setting suggest that the available methods are not capable of producing coherent, logical and consistent results. On top of that, no set methodology to evaluate the accuracy and correctness of the provided explanations is available, nor is a formal methodology to compare the obtained explanations. While the employed techniques provide some sort of insight into the behaviour of the model, experimental results suggest that even though the importance of certain features to the classification is provided, it is relatively easy to produce samples with different labels by perturbing less relevant features. The clear future direction of this field is to establish a formal methodology to compare the xAI methods and to evaluate their quality.

Acknowledgement This research is funded under the Horizon 2020 Ai4Cyber Project, which has received funding from the European Union's Horizon Europe research and innovation programme under grant agreement No. 101070450

References

1. Abadi, M., et al.: Tensorflow: a system for large-scale machine learning. In: 12th {USENIX} Symposium on Operating Systems Design and Implementation ({OSDI} 2016), pp. 265–283 (2016)
2. Capuano, N., Fenza, G., Loia, V., Stanzione, C.: Explainable artificial intelligence in cybersecurity: a survey. IEEE Access **10**, 93575–93600 (2022). https://doi.org/10.1109/ACCESS.2022.3204171
3. Doshi-Velez, F., Kim, B.: Towards a rigorous science of interpretable machine learning (2017). http://arxiv.org/abs/1702.08608
4. Gunning, D., Stefik, M., Choi, J., Miller, T., Stumpf, S., Yang, G.Z.: XAI-Explainable artificial intelligence. Sci. Rob. **4**(37) (2019). https://doi.org/10.1126/scirobotics.aay7120
5. Gurumoorthy, K.S., Dhurandhar, A., Cecchi, G., Aggarwal, C.: Efficient data representation by selecting prototypes with importance weights. In: 2019 IEEE International Conference on Data Mining (ICDM), pp. 260–269. IEEE (2019)
6. Li, L., Jamieson, K., DeSalvo, G., Rostamizadeh, A., Talwalkar, A.: Hyperband: a novel bandit-based approach to hyperparameter optimization. J. Mach. Learn. Res. **18**(1), 6765–6816 (2017)
7. Lundberg, S.M., Lee, S.I.: A unified approach to interpreting model predictions. In: Guyon, I., et al. (eds.) Advances in Neural Information Processing Systems, vol. 30. Curran Associates, Inc. (2017). https://proceedings.neurips.cc/paper_files/paper/2017/file/8a20a8621978632d76c43dfd28b67767-Paper.pdf
8. Molnar, C.: Interpretable Machine Learning (Second Edition) A Guide for Making Black Box Models Explainable. Leanpub (2022). https://leanpub.com/interpretable-machine-learning
9. Mothilal, R.K., Sharma, A., Tan, C.: Explaining machine learning classifiers through diverse counterfactual explanations. In: Proceedings of the 2020 Conference on Fairness, Accountability, and Transparency, pp. 607–617 (2020)
10. Nwakanma, C.I., et al.: Explainable Artificial Intelligence (XAI) for intrusion detection and mitigation in intelligent connected vehicles: a review. Appl. Sci. **13**(3), 1252 (2023). https://doi.org/10.3390/app13031252

11. Pawlicki, M., Kozik, R., Choraś, M.: A survey on neural networks for (cyber-) security and (cyber-) security of neural networks. Neurocomputing **500**, 1075–1087 (2022). https://doi.org/10.1016/j.neucom.2022.06.002
12. Ribeiro, M., Singh, S., Guestrin, C.: Why should i trust you?: explaining the predictions of any classifier. In: 2016 Conference of the North American Chapter of the Association for Computational Linguistics: Demonstrations, San Diego (2016)
13. Sarhan, M., Layeghy, S., Moustafa, N., Portmann, M.: NetFlow datasets for machine learning-based network intrusion detection systems. In: Deze, Z., Huang, H., Hou, R., Rho, S., Chilamkurti, N. (eds.) BDTA/WiCON -2020. LNICST, vol. 371, pp. 117–135. Springer, Cham (2021). https://doi.org/10.1007/978-3-030-72802-1_9
14. Sharafaldin, I., Lashkari, A.H., Ghorbani, A.A.: Toward generating a new intrusion detection dataset and intrusion traffic characterization. ICISSp **1**, 108–116 (2018)
15. Sharma, D.K., Mishra, J., Singh, A., Govil, R., Srivastava, G., Lin, J.C.W.: Explainable artificial intelligence for cybersecurity. Comput. Electr. Eng. **103**, 108356 (2022). https://doi.org/10.1016/j.compeleceng.2022.108356
16. Yan, F., Wen, S., Nepal, S., Paris, C., Xiang, Y.: Explainable machine learning in cybersecurity: a survey. Int. J. Intell. Syst. **37**(12), 12305–12334 (2022). https://doi.org/10.1002/int.23088

Real-Time Deformation of Three-Dimensional Volumetric Models Using Mesh Models

Kamil Szostek

Abstract Medical simulations based directly on CT data are appreciated for their realism in reflecting anatomy. On the other hand, such an approach introduces a number of complications when it comes to the possibilities of constructing interventional simulators. In this paper an approach to CT-based volumetric data deformation is presented using the Obi simulation system and Unity framework. This method allows for easier and faster development, while yielding realistic outcomes built upon established and reliable solution. The method was implemented within the MrTEEmothy simulation system, where deformations of the heart tissues are visible in the simulated ultrasound images, during the transeptal puncture procedure.

1 Introduction

In medical simulations, employing CT-based volumetric data is a means to create a realistic simulation and visualizations. This approach ensures not only anatomical accuracy but also enables training on specific patients using their CT scans. Unfortunately, CT data introduces many limitations and increases processing difficulties. One significant challenge is the implementation of deformation and distortion of tissues. This issue has been discussed in the literature and various methods have been proposed for the implementation of interactive deformations. Torres *et al.* proposed a volume data resampling and constructing a tetrahedral mesh and use it to calculate volume deformations [2, 10]. They have also introduced an interactive deformation of medical 3D scans using a touchscreen interface [11]. In [8] authors use a regular free form deformation structures as a wrapper for a volume deformation, similar to the solution presented in this paper.

In Ultrasound Simulation system, like MrTEEmothy from Medical Simulation Technologies, the CT scans of hearts are used directly to create realistic ultrasound

K. Szostek (✉)
Department of Biocybernetics and Biomedical Engineering, AGH University of Science and Technology, Mickiewicza 30 Av., 30–059 Krakow, Poland
e-mail: szostek@agh.edu.pl

Medical Simulation Technologies, Miechowska St. 5b apt. 4, 30–055 Krakow, Poland

R. Burduk et al. (eds.), *Progress on Pattern Classification, Image Processing and Communications*, Lecture Notes in Networks and Systems 766,
https://doi.org/10.1007/978-3-031-41630-9_6

images of transesophageal echocardiography examinations [3, 7, 9]. This approach allows to experience the real anatomy and pathologies without any interference from a graphic designer. The visualization is based on a 4D data sets, with minimum 10 heart phases and about 512^3 voxels per frame. This data is processed in real time to simulate ultrasound images from transesophagus heart examination.

2 Motivations

Using CT data for multi-modal medical imaging simulations, including ultrasound or fluoroscopy, provides a realistic representation of a patient's tissues, facilitating teaching and skills development. However, when volumetric data is used for simulating surgical procedures, the interactions of the surgical devices and other tools with the volumetric model becomes challenging, as well as its deformation and haptic response.

In this paper an approach to volumetric data deformation is presented, based on an existing solutions available in the Unity Framework [12]. This method offers advantages of easier and faster development, allowing for extensions of existing simulation systems, and delivers realistic results based on well-tested and stable solutions. The volume deformation technique is then used in simulation where the heart tissue reaction is visible on the simulated ultrasound image. Specifically, a deformation of a *septum*, a structure between the atriums of the heart, is presented. This structure is susceptible to deformation during structural heart interventions, i.e. transeptal puncture.

3 Deformation Simulation

In this section the approach to CT data deformation is presented. By leveraging the Obi simulation system available in Unity and authors scripts, the ray marching-based ultrasound simulation was enhanced with the capability to visualize tissue deformations.

3.1 Obi Based Simulation

Obi is a powerful physics simulation engine designed for Unity, a popular framework used for developing 2D and 3D games, simulations, and other interactive applications [6]. With Obi, developers can create realistic and dynamic soft-bodies, cloths, and fluids simulations, which can interact with the environment and other objects in the scene. The engine offers an intuitive and flexible interface, empowering users to adjust physical properties and behaviors to achieve precise control over the performance and appearance of the simulation. Obi features multi-threaded capabilities, however it does not utilize GPU for computation purposes.

The Obi approach is to simulate all physics using particles that can interact with one another and be affected by other objects. This methodology implements the solution proposed in [5], where authors shows a technique for resampling a vertex model into particles with defined constraints. The physics calculations are performed on these particles, and the results are then mapped to the original vertex mesh. In this work, the Obi's particles are mapped onto the CT-based volumetric model.

The physical simulation in Obi system is controlled by a wide range of adjustable parameters to cater to specific requirements. These parameters can be fine-tuned at different levels, including individual particle properties such as mass, plasticity, and size, as well as global settings for the entire soft-body object and the solver. The solver is a core processing module, where simulation details can be defined, including environment parameters like gravity, damping, collisions and the number of iterations for constraint calculations. The correct set up of the solver and the particle-based object's details is essential in achieving accurate simulation results for the selected soft body, i.e. heart tissues. Using the Obi's editor, the user can precisely set the physical parameters of objects.

3.2 3D Volume Deformation

The first step of volumetric model deformation was to create a triangle based 3D models from the selected tissue of the volume model. This segmentation process was performed manually using the 3D Slicer software [1], utilizing the methods proposed in [4]. The outcome of the partial segmentation (only septum) is presented on Fig. 1a.

The segmented model was imported into Unity framework, simplified and transformed into a soft-body "particle blueprint". This blueprint contains particles information required to conduct a simulation. The Obi plugin was used to automatically populate particles over the mesh, using a predefined density (Fig. 1b). This step is one of the crucial part in ensuring both the quality and performance of the final simulation.

With the particle model in place, the next step requires a solver configuration. Depending on the requirements and complexity of the simulation system, some of the constraints iterators can have a reduced number of steps or be disabled altogether. The parameters like damping, gravity, sleep threshold and number of collision's iterations must be adjusted.

After setting up the scene with volume models and Obi particles, the simulation can be performed. To reflect the deformations of the particle structure, it is necessary to map them to the voxel data in rendering algorithm. The following approach was implemented: the positions of the particles and their translated location, are copied to the GPU buffers and utilized in a compute shader. This shader creates a 3D texture,

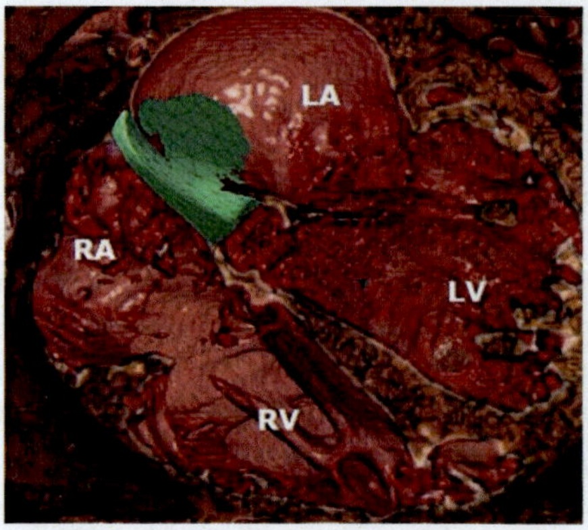

(a) The *septum* tissue segmentation, represented as a green mesh (L - left, R - Right, A - atrium, V - ventricle).

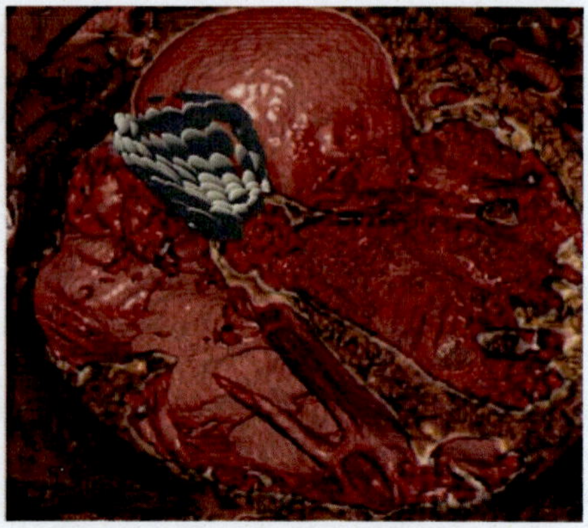

(b) Particles used for calculations.

Fig. 1 Rendering of the CT scan data, heart model, with *septum* vertex model and particle representation. Volume rendering is performed using author's custom method, implemented in Unity Framework.

where each voxel V at position x, y, z contains a vector of its deformation. The vectors are calculated as a weighted sum of all particle transitions (Eq. 1):

$$V'_{x,y,z} = \sum_{i=1}^{N} -(\boldsymbol{T_i} - \boldsymbol{S_i})e^{-H\|V_{x,y,z} - S_i\|^2} \tag{1}$$

$V_{x,y,z}$ - original sampling source at x, y, z
N - number of particles,
$\boldsymbol{S}$ - position of source particle,
$\boldsymbol{T}$ - position of current particle,
H - a distance scaling factor.

In this approach the performance is dependent on the number of particles and the size of the generated proxy texture. However, it can be optimized by adding an influence distance limit.

The resulting 3D texture contains a deformation map that is applied during the rendering of the simulated ultrasound image. It can be also used for 3D volume visualization of the heart model.

The number of particles generated for the *septum* model was 115, 417, 965 and 1546 (Fig. 2). The deformation 3D texture size was 64^3, 128^3, 196^3 and 256^3 voxels. The heart model was a CT data set of $512 \times 512 \times 640$ voxels. The resulting ultrasound image was a 512×512 pixels. The visualization was done using a Intel(R) Core(TM) i7-8700 CPU, 3.20 GHz and NVIDIA GeForce RTX 2060 GPU. The average framerate was 30 fps (including US simulation and deformations). The performance results are presented on the Fig. 3 and Table 1 (Fig. 4).

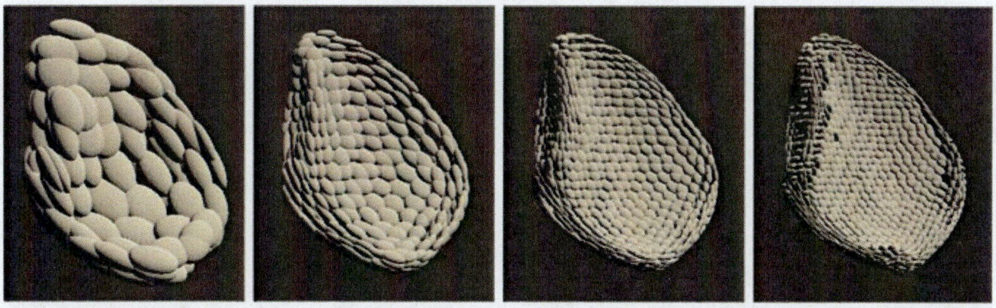

(a) 115 (b) 417 (c) 965 (d) 1546

Fig. 2 Different numbers of particles generated for the *septum* mesh.

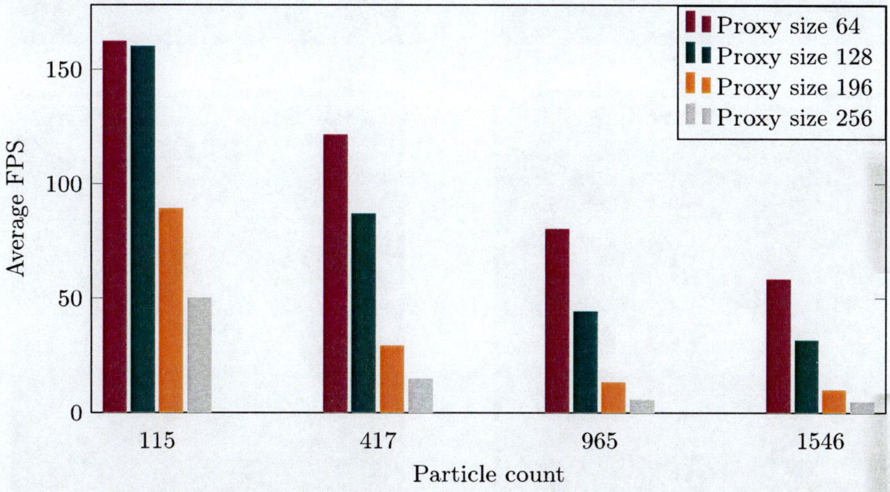

Fig. 3 Average number of frames per seconds (FPS) for different configurations.

Table 1 Average number of frames per seconds for different proxy sizes and particles counts.

	Proxy Size			
Particle count	**64**	**128**	**196**	**256**
115	162	160	88.9	50.1
417	121.1	86.8	28.9	14.7
965	80	44.1	13.1	5.7
1546	58	31.5	9.7	4.8

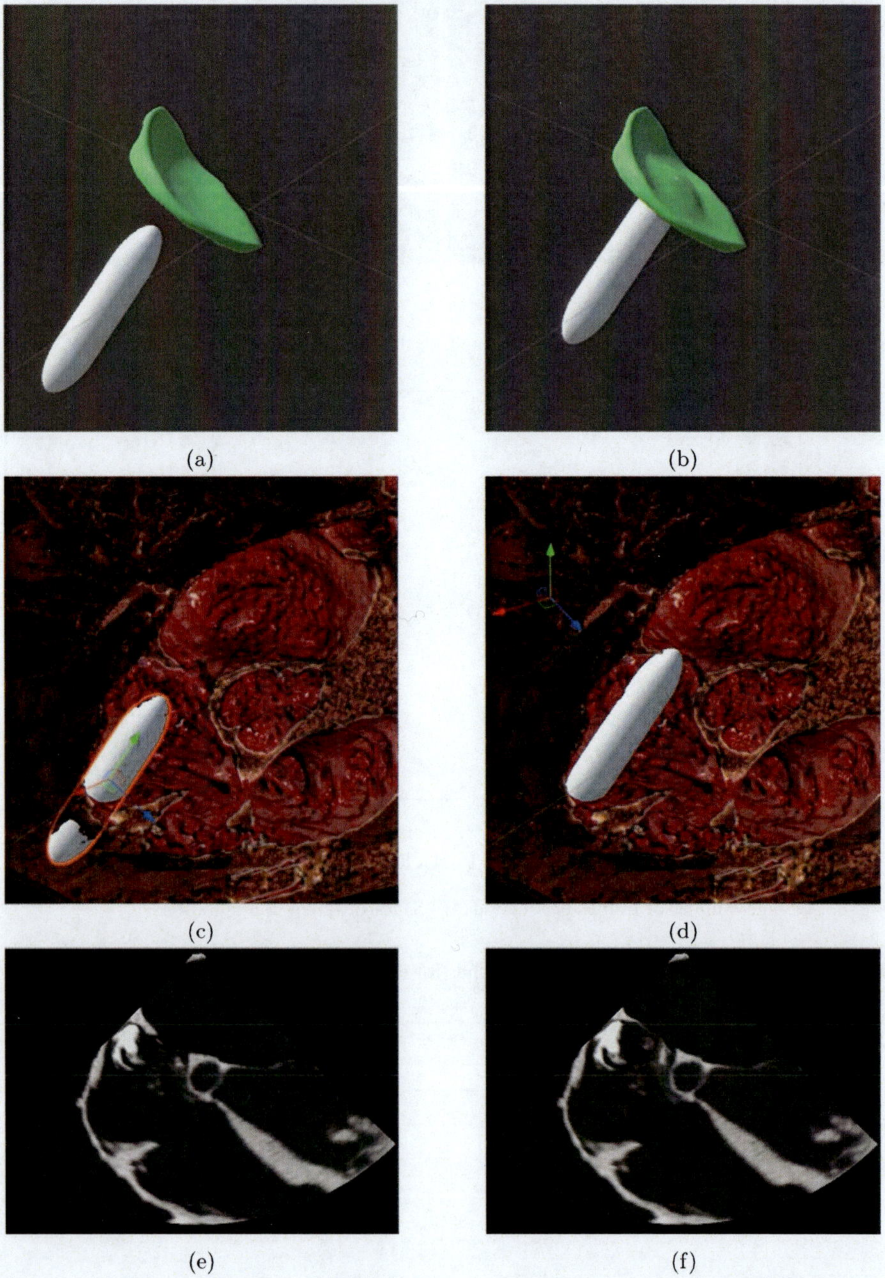

Fig. 4 The rendering of the *sptum* mesh (a, b), CT data (c, d) and simulated ultrasound images (e, f). The right side has no deformation, left side demonstrates the influence of a custom object on the tissue.

4 Results and Discussion

In this work, a solution for a volume data deformation is presented. The Unity framework with Obi extension was used to implement complex soft body physics, i.e. elastic and plastic deformation, and map it onto CT data. This method provides an easy way to extend existing ultrasound simulation platform with device interactions, including not only rigid bodies, but also other types of physical object supported by Obi extensions, such as ropes, rods, clothes or fluids. Furthermore, this solutions offers a straightforward approach to implementing haptic feedback if needed, as the Obi extensions provides force data for each particle or body used in simulation.

The performance and quality of the simulation are influenced by the size of the proxy volumes and the number of particles. Table 1 shows that the best performance is achieved with small proxy volumes and a limited number of particles. However, this approach decreased the level of detail that can be visualized.

One of the limitations of presented method is the manual segmentation process required to create the 3D model of the heart. Although segmentation methods are well-known, they are often time-consuming and require a high level of expertise. Therefore, automating the segmentation process would be desirable to reduce the time and effort required.

Another limitation is the lack of details in the resampled models. The sampled volume points are limited by the resolution of the mesh model, which can result in a loss of detail in the 3D volume model. This may impact the accuracy of the simulation, particularly for small or intricate structures. Increasing the size of the resampled models have significant impact on the performance. However, if the physical simulation is not required for the whole model all the time, it could be divided into smaller parts or with different sampling rate, increasing the quality and the simulation performance.

Finally, the lack of GPU computing in the Obi plugin can also be considered as a limitation of presented method. Although presented method is capable to create real-time simulation, it still requires significant computational resources. Using GPU computing could further accelerate the simulation process, allowing for even faster and more efficient simulations.

Acknowledgment The financial support was provided by the Regional Operational Programme for the Małopolska Region 2014–2020, grant number RPMP.01.02.01-12-0027/19-00, title "Digital Transseptal Puncture Simulator". The project was also partially funded by the Doctoral School of University of Science and Technology, Cracow, Poland, as a part of the Industrial Doctoral Program "Selected phenomena simulation using ray trace technique".

References

1. Fedorov, A., et al.: 3D slicer as an image computing platform for the quantitative imaging network. Magn. Reson. Imaging **30**(9), 1323–1241 (2012). https://doi.org/10.1016/j.mri.2012.05.001. https://www.slicer.org/
2. Gascon, J., Espadero, J.M., Perez, A.G., Torres, R., Otaduy, M.A.: Fast deformation of volume data using tetrahedral mesh rasterization. In: Proceedings of the 12th ACM SIGGRAPH/Eurographics Symposium on Computer Animation, SCA 2013, pp. 181-185. Association for Computing Machinery, New York (2013). https://doi.org/10.1145/2485895.2485917
3. Kempny, A., Piórkowski, A.: CT2TEE - a novel, internet-based simulator of transoesophageal echocardiography in congenital heart disease. Kardiologia Polska **68**(3), 374–379 (2010)
4. Lasek, J.: The impact of data preprocessing on the accuracy of CNN-based heart segmentation. In: Choraś, M., et al. (eds.) CORES/IP&C/ACS -2021. LNNS, vol. 255, pp. 173–180. Springer, Cham (2022). https://doi.org/10.1007/978-3-030-81523-3_17
5. Macklin, M., Müller, M., Chentanez, N., Kim, T.Y.: Unified particle physics for real-time applications. ACM Trans. Graph. **33**(4) (2014). https://doi.org/10.1145/2601097.2601152
6. Mendez, J.M.: Obi: Unified particle physics for unity (2023). http://obi.virtualmethodstudio.com
7. Piórkowski, A., Kempny, A.: The transesophageal echocardiography simulator based on computed tomography images. IEEE Trans. Biomed. Eng. **60**(2), 292–299 (2013)
8. Prados, F., León Salas, A., Torres, J.: Haptic interaction with elastic volumetric structures. Int. J. Creat. Interfaces Comput. Graph. (IJCICG) **3**, 63–73 (2012). https://doi.org/10.4018/jcicg.2012010105
9. Szostek, K., Piórkowski, A., Kempny, A., Banyś, R., Gackowski, A.: Using computed tomography images for a heart modeling. J. Med. Inf. Technol. **19**, 75–84 (2012)
10. Torres, R., Rodríguez, A., Espadero, J.M., Otaduy, M.A.: High-resolution interaction with corotational coarsening models. ACM Trans. Graph. **35**(6) (2016). https://doi.org/10.1145/2980179.2982414
11. Torres, R., Rodríguez, A., Otaduy, M.: Hands-on deformation of volumetric anatomical images on a touchscreen. Appl. Sci. **11**(20) (2021). https://doi.org/10.3390/app11209502
12. Unity Technologies: Unity Real-Time Development Platform | 3D, 2D, VR & AR Engine (2023). https://unity.com/. Accessed 24 Apr 2023

Hyperparameters Optimization Using GridSearchCV Method for TinyML Models

Tobiasz Puślecki and Krzysztof Walkowiak

Abstract The anticipated expansion of TinyML-related technologies will require methods for efficient ML model development. In this paper, we present a method for obtaining a set of ML models for TinyML systems that satisfies the assumption that a more efficient model is also more complex and therefore consumes more energy. The results show that our method is capable of providing numerous diversified sets of ML models.

Keywords TinyML · hyperparameters · optimization

1 Introduction

Classical machine learning (ML) models require quite large computational and memory resources. To apply ML methods to resource-constrained platforms, special techniques would have to be used to design and implement simplified models. This niche is filled by TinyML. The term is a combination of hardware, software, and applications that require very low power consumption (typically in the milliwatt range and below) by a battery-powered device that cannot handle complex models [13]. ML models running on TinyML devices must therefore be optimized for size and performance.

A less computationally advanced model will take less time to recalculate, making the inference process shorter. This implies lower power consumption [7, 11]. The difference in architecture and choice of hyperparameters may imply differences in the efficiency and complexity of models in certain use cases [3, 6, 9]. There are use cases where it is possible to find models for which a larger model is more efficient [15].

T. Puślecki (✉) · K. Walkowiak
Department of Systems and Computer Networks, Wrocław University of Science and Technology, Wrocław, Poland
e-mail: tobiasz.puslecki@pwr.edu.pl

K. Walkowiak
e-mail: krzysztof.walkowiak@pwr.edu.pl

R. Burduk et al. (eds.), *Progress on Pattern Classification, Image Processing and Communications*, Lecture Notes in Networks and Systems 766,
https://doi.org/10.1007/978-3-031-41630-9_7

In our previous paper [12], we showed that selecting a set of models appropriately (where each successive model is slightly worse but more energy-efficient) and changing them allows for additional operational time at the cost of slightly worse prediction quality. The difficulty was to create a set of models meeting the mentioned assumption - we had to manually search for combinations of hyperparameters for a given architecture, so we plan to automate the procedure of creating sets of models.

In this article, we present a method for obtaining a set of ML models for TinyML systems that satisfies the aforementioned assumption, that is, a computationally more complex model is both more accurate but consumes more energy than a less complex and less accurate model. The main contribution and novelty of this work is a new algorithm for creating this type of model set, which has not yet been developed.

The rest of the paper is organized as follows. Section 2 describes the GridSearchCV method. Section 3 describes `TGrid` algorithm. Section 4 describes conducted experiments. Section 5 concludes the work.

2 GridSearchCV Method

Choosing the suitable type of classifier and the relevant set of hyperparameters has a key impact on the effectiveness of the overall classification process. In addition, an important aspect is the choice of metrics, and the simplest of these is the metric of classification accuracy, measured as the percentage of correctly recognized examples.

To avoid overfitting, the classifier evaluation should be carried out using data that has not previously been used for training. In the method presented here, a cross-validation technique is used, which involves dividing the training set into K parts, then rejecting one part in turn, teaching the classifier on the remaining parts, and evaluating the rejected part. This sequence of steps is repeated K times (each time rejecting a different part). Finally, the results are averaged with K evaluations [2].

After choosing the algorithm, metrics, and data set, the hyperparameters space for optimization must be defined. Note that each additional hyperparameter increases the time to search the space, expanding it by another dimension. It now remains to pass the hyperparameters space in the form of a dictionary as the param_grid argument to the GridSearchCV method. It is worth noting that for non-linear spaces (or those at which a single hyperparameter may have a minor effect on the results) it is better to use randomized search. RandomizedSearchCV (GridSearchCV alternative) takes probability distributions of hyperparameters as its main argument (param_distribution). Properly defined distributions allow to search the space faster over a larger range of hyperparameters and their values. This approach works well when there are a large number of hyperparameters to be optimized, or when you need to know the range of hyperparameters that will be used as arguments to the GridSearchCV method [4].

3 TGrid Algorithm

For our research, it is crucial to obtain a set of models that satisfy the assumption (labeled MA) that the larger the model, the more effective it is, and the longer it takes to calculate (which implies higher energy consumption). Assuming that we have the chosen algorithm (classifier/regressor), evaluation metric, and learning dataset, it remains to find a set of hyperparameters. Manually searching for the optimal set is time-consuming and annoying, so it is necessary to perform the process of searching the hyperparameter space in an automated way - here using the GridSearchCV method described in the previous section. Depending on the chosen method of optimization of hyperparameters of classification models in the learning process, we will obtain a certain subset B of hyperparameter combinations (from which card(B) models can be produced) from the chosen space A (for GridSearchCV the subset is equal to the whole space A=B, for RandomizedSearchCV B$\subset$A).

The result of GridSearchCV is an array of results containing, among other things, the average `ACC` (accuracy) value of the given evaluation metric from K evaluations. Each model is characterized by the number of floating-point operations (`FLOP`) in a single inference operation. We take this as a measure of the computational complexity of the model.

Algorithm 1 presents the pseudocode of the described procedure. We assume that each model is characterized by a triplet `(model_Tag, ACC, FLOP)`, where `modelTag` is a unique identifier of the model (*Line 1*). We sort the set of such triplets by the value of `ACC`, which results in a `TL` list (*Line 3*). Then, iterating through the `TL` list in a loop, as long as the `TL` is non-empty (*Line 5*) we take the first element of the list (here the best model; the i-th iteration - the best model with `model_tag` i, in other words, we change the best model in each iteration, and each iteration will contain min. 1 element) and we add it to the `partial_list`. We then add more triplets to the `partial_list` so that successive `FLOP` values form a descending sequence (*Lines 7–13*). We remove the first list element from the `TL` (*Line 15*). We add the `partial_list` to the `partial_lists` (*Line 14*). We clean the `partial_list` (*Line 6*). After going through the entire `TL` as a result of the algorithm, we have a list of `partial_lists` from which a minimum of one combination satisfying the MA assumption can be selected (*Line 17*).

Depending on the needs or a particular use case, one can choose one of the lists returned by the `TGrid` method. For example, it can be the longest list, which will provide many diversified models, however, the strategy for selecting a specific list is a separate problem and beyond the scope of this article.

To summarize, the proposed algorithm returns sets of hyperparameters for the models. In each set, the models are ranked from best to worst and from most complex to least complex.

Algorithm 1: `TGrid` algorithm

```
triplets ← list of (model_tag, ACC, FLOP in triplets) triplets;
Procedure TGrid(triplets)
    sort(triplets).by(ACC);
    partial_lists ← [];
    while triplets not empty do
        partial_list ← [];
        previous ← infinity;
        foreach model_tag, ACC, FLOP in triplets do
            current ← FLOP;
            if (current < previous) then
                previous ← current;
                partial_list.append((model_tag, ACC, FLOP));
        end
        partial_lists.append(partial_list);
        /* Pop first element                                    */
        triplets.pop(0);
    end
    return partial_lists
```

4 Experiments

A set of handwritten digits MNIST [8] has been chosen for experiments. We will consider a multi-class classification problem (10 classes for 10 digits from 0 to 9). Using the TensorFlow library [1] and Keras [5], we built a model with the architecture described in Table 1. Table 2 shows the parameters passed as arguments to the GridSearchCV method, where:

- **conv_layer1** and **conv_layer2** are possible hyperparameters values for convolutional layers,
- **dense_layers** are possible hyperparameters values for dense layers,
- **epochs** are possible number of epochs,
- **batch_size** are possible values of batch size,
- **cv** specifies the number of folds in a (Stratified)KFold.

The hyperparameters of this model are processed using the GridSearchCV method from the Scikit-learn [10] library. The FLOP value is calculated using the Keras-flops library [14].

Figure 1 shows in the graph on the left-hand side the value of the accuracy metric for each model (highlighted by `model_tag` number). The graph on the right-hand side shows the value of the `FLOP` metric for each model. The number of all models shown is 192. It can be seen that, according to the assumption presented earlier, the sequence of successive values of the accuracy metric is decreasing. However, the sequence of successive values of the `FLOP` metric is not a monotonic sequence, so it is necessary to select models in such a way that each successive slightly less accurate model is at the same time less computationally demanding.

Table 1 Architecture of studied neural network

Layer	Description
Conv2D	Mandatory
MaxPooling2D	Mandatory
Conv2D	Optional
MaxPooling2D	Optional
Flatten	Mandatory
Dense	Mandatory
Dense	Optional
Dense	Mandatory

Table 2 Parameters passed as arguments to GridSearchCV method

Params—possible hyperparameters values	
conv_layer1	[4,3,2,2], [4,5,2,2], [8,3,2,2], [8,5,2,2]; [16,3,2,2], [16,5,2,2]
conv_layer2	[], [2,3,2,2], [4,3,2,2], [8,3,2,2]
dense_layers	[50], [50,20]
epochs	5, 10
batch_size	64, 128
cv	5

As an example, here is the set resulting from iteration number 3. Figure 2 shows in the graph on the left-hand side the value of the accuracy metric for each model in set number 3 (as an example). The graph on the right-hand side shows the value of the `FLOP` metric for the model in set number 3. Set number 3 contains 12 different models. Since the selection of models was made using the presented Algorithm 1, the sequence of successive values of the `FLOP` metric is a descending sequence.

Figure 3 shows in the graph on the left-hand side the value of the accuracy metric for each model in set number 111 (as an example). The graph on the right-hand side shows the value of the `FLOP` metric for the model in set number 111. Set number 111 contains 8 different models. The accuracy metric values for set number 3 and set number 111 are similar, but the `FLOP` metric values form a steeper graph for set number 111. This means that by switching to a less efficient model for set number 111, more energy can be saved than for set number 3.

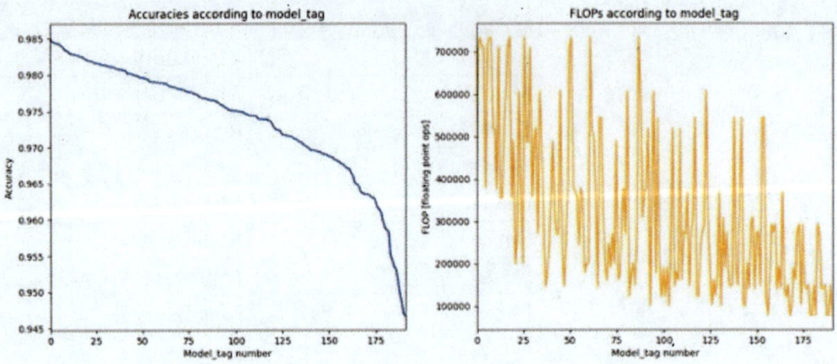

Fig. 1 Accuracies and FLOPs according to model_tag

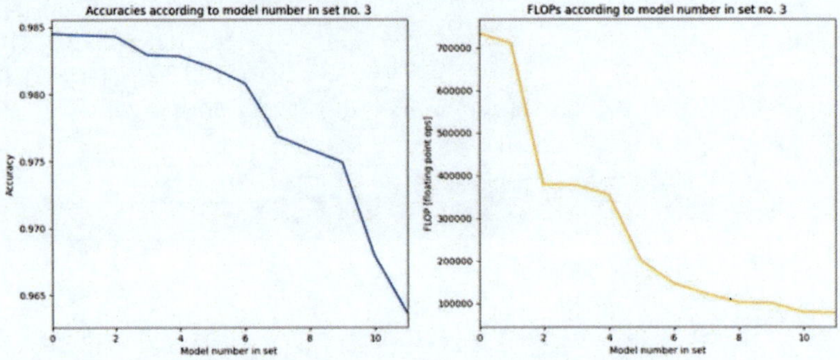

Fig. 2 Accuracies and FLOPs according to model number in set no. 3

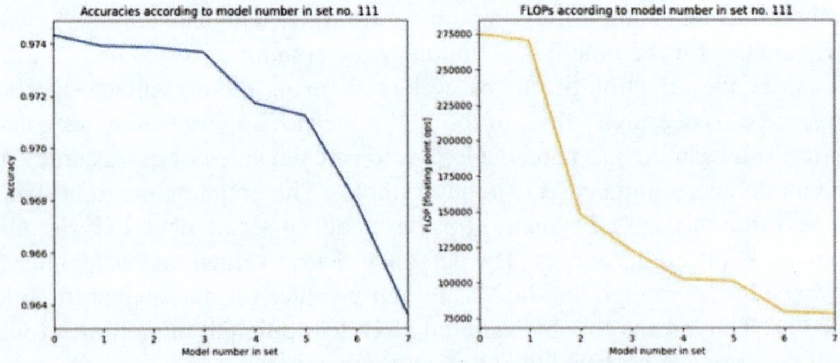

Fig. 3 Accuracies and FLOPs according to model number in set no. 111

5 Conclusions

In this paper, we have presented a method for producing sets of different ML models that can allow an attempt to make a trade-off between accuracy and power consumption in TinyML systems. In other words, using these models it will be possible to optimize operation time in TinyML-based applications by balancing the quality/energy consumption trade-off.

Future experiments are planned with the RandomizedSearchCV method, as well as experiments combining the RandomizedSearchCV and GridSearchCV methods - the first can be used to determine the neighborhood of the local maximum, and the second to search that neighborhood. We also plan to create our own FLOP counting library for models in TensorFlow Lite format.

References

1. Abadi, M., et al..: TensorFlow: large-scale machine learning on heterogeneous systems (2015). https://www.tensorflow.org/
2. Arlot, S., Celisse, A.: A survey of cross-validation procedures for model selection. Stat. Surv. **4**, 40–79 (2010). https://doi.org/10.1214/09-SS054
3. Bechtel, M., Weng, Q., Yun, H.: Deeppicarmicro: applying tinyml to autonomous cyber physical systems. In: 2022 IEEE 28th International Conference on Embedded and Real-Time Computing Systems and Applications (RTCSA), pp. 120–127 (2022). https://doi.org/10.1109/RTCSA55878.2022.00019
4. Bergstra, J., Bengio, Y.: Random search for hyper-parameter optimization. J. Mach. Learn. Res. **13**, 281–305 (2012)
5. Chollet, F., et al.: Keras. https://keras.io (2015)
6. Chowdhury, A.A., Hossen, M.A., Azam, M.A., Rahman, M.H.: Deepqgho: quantized greedy hyperparameter optimization in deep neural networks for on-the-fly learning. IEEE Access **10**, 6407–6416 (2022). https://doi.org/10.1109/ACCESS.2022.3141781
7. Giordano, M., Piccinelli, L., Magno, M.: Survey and comparison of milliwatts micro controllers for tiny machine learning at the edge. In: 2022 IEEE 4th International Conference on Artificial Intelligence Circuits and Systems (AICAS), pp. 94–97 (2022). https://doi.org/10.1109/AICAS54282.2022.9870017
8. LeCun, Y., Cortes, C.: MNIST handwritten digit database. http://yann.lecun.com/exdb/mnist/ (2010)
9. Lin, J., Chen, W.M., Lin, Y., Gan, C., Han, S.: Mcunet: tiny deep learning on IoT devices. Adv. Neural Inf. Process. Syst. **33** (2020)
10. Pedregosa, F., et al.: Scikit-learn: machine learning in python. J. Mach. Learn. Res. **12**, 2825–2830 (2011)
11. Piechocki, M., Kraft, M., Pajchrowski, T., Aszkowski, P., Pieczynski, D.: Efficient people counting in thermal images: the benchmark of resource-constrained hardware. IEEE Access **10**, 124,835–124,847 (2022). https://doi.org/10.1109/ACCESS.2022.3225233
12. Puślecki, T., Walkowiak, K.: On the selection of a machine learning model in tinyml devices - preliminary study. In: 4th Polish Conference on Artificial Intelligence (PP-RAI) (2023)
13. TinyML Foundation Organization Homepage. https://www.tinyml.org. Accessed 25 Feb 2022
14. Tokusumi: Keras-flops. https://github.com/tokusumi/keras-flops (2020)
15. Zhang, Y., Wijerathne, D., Li, Z., Mitra, T.: Power-performance characterization of tinyml systems. In: 2022 IEEE 40th International Conference on Computer Design (ICCD), pp. 644–651 (2022). https://doi.org/10.1109/ICCD56317.2022.00099

Analysis of the Possibility to Employ Relationship Between the Problem Complexity and the Classification Quality as Model Optimization Proxy

Joanna Komorniczak, Paweł Ksieniewicz, and Michał Woźniak

Abstract Bulk construction of pattern classifiers, whether for optimizing input data configurations or method hyperparameters, is a computationally highly complex task. The main problem is the prediction quality evaluation function based on estimation using the selected experimental protocol. In the case of iterative optimization algorithms, such an evaluation is computationally-intensive, runs in each iteration, and requires a separate data partition for quality estimation. So-called proxy models may be alternative solutions, which estimate classifier quality on data characteristics without the need to train the prediction model. There are some premises that the problem complexity measures can be used for this purpose. However, this paper negatively verifies this hypothesis – confirming the predictive potential of evaluating the effectiveness of models by complexity measures but also showing a relatively large measurement error in direct relation between quality metric and *proxy* measure.

Keywords Data complexity · Imbalanced data · Optimization · Data resampling

1 Introduction

The pattern classifier quality depends on many factors, i.e., a training data selection or model hyperparameterization, that makes the calibration or fine-tuning – leading to the maximization of the generalization ability – an extremely computationally costly task, exceeding the cost of constructing the classifier, which is especially noticeable with more complex models. A potential solution, which has already been initially verified by the literature [11], is using so-called optimization proxies, where the verification of the pre-processed input data potential is not based on the target model but is estimated based on predetermined predictors remaining in a specific known statistical relationship to the classifier efficiency.

J. Komorniczak (✉) · P. Ksieniewicz · M. Woźniak
Department of Systems and Computer Networks, Wrocław University of Science and Technology, Wrocław, Poland
e-mail: joanna.komorniczak@pwr.edu.pl

R. Burduk et al. (eds.), *Progress on Pattern Classification, Image Processing and Communications*, Lecture Notes in Networks and Systems 766,
https://doi.org/10.1007/978-3-031-41630-9_8

Calibrating a classifier usually goes beyond a simple review of its hyperparameters – focusing on the pre-processing phase – opening the broad door to multi-criteria optimization strategies. This is particularly interesting in the case of problems characterized by class imbalance, i.e., the disproportion between the sizes of categories distinguished by the system. Default strategies for mitigating imbalanced problems fall into three main categories: (a) built-in methods – capable of effectively inducing an imbalanced pool of training observations, (b) pre-processing methods – modifying this pool so that the model can recognize correctly despite the original lack of properties enabling the proper understanding of minority classes and (c) hybrid methods – most often combining pre-processing with the construction of a vast pool of models cooperating under a given or learned integration rule.

In such cases, pre-processing algorithms strongly depend on the randomness, such as *Random Over-* and *Undersampling* or *SMOTE*. In those cases, the random starting solution is burdened with an inevitable error, which could be reduced by extending the selected pre-processing method with an optimization procedure that maximizes the selected pool of metrics sensitive to the specificity of the task. However, it should be noted that each individual in the population used in such a procedure could be appropriately assessed in terms of its potential only after a full, time-consuming training of the model. Even apart from the fact that the classification method itself may be additionally burdened with its non-determinism of error estimation, it is not easy to consider it possible and achievable to use such a strategy – for example – in the search for the optimal configuration of input data for a deep neural network.

1.1 Related Works

The solution to this problem may be identifying factors statistically related to the model's input data and error level. Natural candidates considered for this role by the literature are measures of the complexity of recognition problems. The scope of data complexity measures applications, according to the available literature, is relatively broad, ranging from applications in signal data [14], through the dominant application in meta-learning [20], extending the context of difficult data analysis [7], or applications in supporting the classification models of imbalanced data classification [1].

Assessing the problem's difficulty often appears in the literature as a tool to extend the analysis of results. This applies to research in the field of digital image processing, when assessing the impact of the thermal image scale on the difficulty of the classification task solved by LSTM networks [19], and to the issue of model explainability, when assessing the impact of increasing the explainability of the model on changes in problem difficulty was measured by the $F1v$, $N3$ and $L1$ measures [7]. There is also a noticeable trend taking into account changes in the complexity of the problems after the resampling phase for the imbalanced data classification [12].

The primary scope of research on imbalanced data classification is based on evaluating the relationship between the resampled training set and the test set [5]. As part of the research on the correlation of measures with the quality of recognition in an imbalanced environment, [1] proposed versions of metrics adapted to the disturbed prior distribution. Practical applications in such environments allow determining the size of the synthetic set in oversampling or prior recommendation of the most appropriate recognition methods for a given problem [4]. It is worth mentioning [21] where the authors evaluated the difficulty of simultaneous class imbalance and overlapping.

As an issue integrating all areas of research in data complexity measures, the study of the relationship between model quality and classification task difficulty can be distinguished. The base analyses, conducted both on small collections of benchmark data [3], on dimensionality reduction [17] and imbalanced sets [1] initially confirm the possibility of making it plausible. This provides the basis for conducting research on selecting preprocessing models and optimizing their configuration, an example of which may be the work of [2].

A set of data complexity measures is presented in [16]. This work describes 22 measures, divided into six categories and implemented in *ECoL* package for *R*, *DCoL* package for *C++* and in *problexity* package for *Python* language [10]. All the measures are presented in Table 1. The original implementation of measures with a single asterisk (*) was slightly modified by authors of *ECoL* package, and for those with a double asterisk (**), authors have selected the required parameters.

1.2 Contribution

A change in the values of problem complexity measures due to oversampling procedure has already been observed [11] in the previous study of this topic. The original publication also describes the correlation between specific problem difficulty measures and recognition quality, mainly focusing on the correlation of their relative values. It was observed that the difficulty of the classification problem and the classification quality metrics changed after subjecting the dataset to the oversampling procedure compared to its base value – without data modification.

This work continues the previously conducted analysis, extending it beyond pure correlation analysis to use a combination of the considered metrics as an indirect model for estimating the classification error. In the original analysis, six basic oversampling algorithms, all 22 measures of classification difficulty described by Lorena [16], and five base classifiers were evaluated. A negative correlation is observed for the F1 and N2 measures and the balanced accuracy score metric for the MLP classifier. Therefore, there is a noticeable predictive potential here, which would be worth verifying experimentally in a specific recognition context.

It should be mentioned, nonetheless, that most guided approaches employ some heuristics. Some authors approach preparing imbalanced data as an optimization task to avoid applying any "*rule of a thumb.*" For instance, an evolutionary algorithm for the undersampling was suggested [9]. Undersampling was also handled as a multi-

Table 1 Utilized measures of classification problem complexity assessment

CATEGORY	MEASURE	SYMBOL
Feature-based	Maximum Fisher's discriminant ratio	$F1$*
	Directional vector maximum Fisher's discriminant ratio	$F1v$
	Volume of overlapping region	$F2$
	Maximum individual feature efficiency	$F3$
	Collectove feature efficiency	$F4$
Linearity	Sum of error distance by linear programming	$L1$
	Error rate of linear classifier	$L2$
	Non linearity of linear classifier	$L3$**
Neighborhood	Fraction of borderline points	$N1$
	Ratio of intra/extra class NN distance	$N2$
	Error rate of NN classifier	$N3$
	Non linearity of NN classifier	$N4$**
	Fraction of hyperspheres covering data	$T1$*
	Local set average cardinality	LSC
Network	Density	$density$**
	Clustering Coefficient	$ClsCoef$**
	Hubs	$Hubs$**
Dimensionality	Average number of features per points	$T2$*
	Average number of PCA dimensions per points	$T3$
	Ratio of the PCA dimension to the original dimension	$T4$
Class imbalance	Entropy of class proportions	$C1$
	Imbalance ratio	$C2$

criteria optimization problem [6], and metaheuristic algorithms [15] enabled several directed oversampling methods.

2 Purposes of the Experiment

We proposed two computer experiments to evaluate the potential of problem complexity measures: (i) optimization based on a single criterion and (ii) evaluation of the relationship between difficulty and quality with a variable sampling of the problem evidence available for the training procedure.

2.1 Experiment 1: Single Criterion Optimization

The correlation between the classification complexity measures and quality could become helpful in optimization procedures. We could potentially use a less computationally costly substitute criterion – the measure of problem complexity – rather than performing an evaluation using classification models during the optimization procedure.

To verify this hypothesis, a genetic algorithm based on *Random Oversampling* was designed to perform a dataset oversampling while optimizing the measures of problem complexity to a particular criterion value. The algorithm multiplied the instances of the minority class to achieve a certain proportion between the classes set by an input parameter. The measure of complexity was optimized to the target value.

Experimental Setup. The experimental analysis was performed for six imbalanced datasets listed below (IR stands for imbalance ratio, i.e., the proportion of minority classed in the dataset, and #n denotes the number of samples in the dataset).

- `glass-0-4_vs_5` – #n=92, IR=0.1,
- `ecoli-0-6-7_vs_5` – #n=220, IR=0.09,
- `ecoli-0-1_vs_2-3-5` – #n=244, IR=0.1,
- `glass-0-1-4-6_vs_2` – #n=205, IR=0.08,
- `ecoli-0-3-4-6_vs_5` – #n=205, IR=0.1,
- `glass-0-1-6_vs_5` – #n=184, IR=0.05.

The configuration of experiments was performed to analyze two $F1$ and $N2$ criteria with target in range 0.2-0.8 and imbalance ratio – denoting the share of the original minority class – in range 0.1-0.9. From the static configuration, crossover ratio was set at 0.05, mutation ratio at 0.01, population size with 100 specimens, where optimization was conducted over 200 iterations.

The experimental protocol based on the Stratified 5×2 Cross-Validation. Each training set was subjected to the resampling procedure, then the quality of MLP classifier recognition was evaluated on the testing set. The MLP classifier implementation was taken from the scikit-learn [18] library. This procedure was repeated ten times for each dataset.

2.2 *Experiment 2: Dependency Research for Resampling Methods*

Optimizing a single criterion did not bring benefits to the recognition quality. Another hypothesis has emerged that there may be some more complex relationship between the difficulty measures and the classification quality in the resampling procedure.

Another experiment was carried out to investigate the effect of resampling methods on the difficulty measures in more detail. A single binary synthetic dataset was generated according to the standard Madelon procedure proposed by Guyon [8]. The dataset consisted of 1000 instances, of which 10% were negative and 90% positive.

Experimental Setup. The behavior of 9 resampling methods was tested – 5 oversampling (*Random Oversampler* – ROS, SMOTE, *SVM SMOTE* – SVMS, *Borderline SMOTE* – BLS and *ADASYN* – AS) and 4 undersampling (*Random Undersampler* – RUS, *Cluster Centroids* – CC, *Condensed Nearest NeighBour* – CNN and *One Sided Selection* – OSS), the implementation of which came from the imbalanced learn [13] library.

Each tested algorithm was executed 1000 times, resampling to random class proportions. Values ranging from the initial class ratio to the total balance were drawn. It was not possible to set the target ratio for two of the tested undersampling algorithms. For these cases the default hyperparameterization was used. The problem's complexity and classification quality was examined using the MLP algorithm, initiated with the same weights for each new perturbation of the set. Classification quality was estimated with the Stratified 5×2 Cross-Validation. The implementation of the MLP classifier and regressor, later used to analyze the results.

3 Experimental Evaluation

Let us present the course of the planned experiments and interprets the obtained results.

3.1 *Experiment 1: Single Criterion Optimization*

This section summarizes the results of the performed experiment regarding the evaluation of optimization ability and the quality of the recognition model fitted on resampled data.

Evaluation of Optimization Ability. Figure 1 shows the criterion value after optimization to the target value and the optimization error – criterion value after optimization minus the target value. The summary aimed to verify the genetic algorithm's effectiveness and examine which values are achieved in the optimization process.

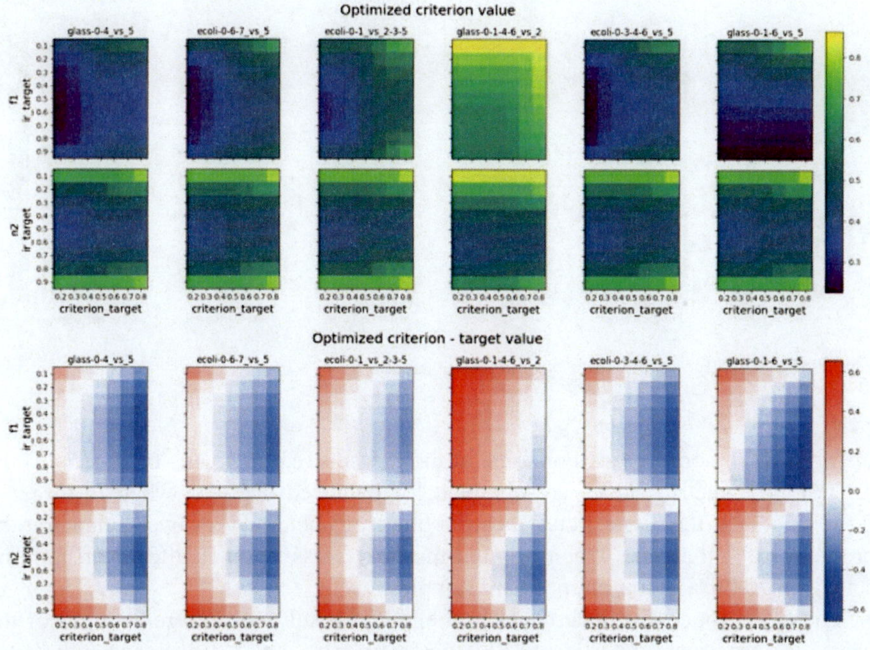

Fig. 1 Value of optimized criterion and its difference form the target

The combinations of parameters for which the achieved criterion value was higher than the target one are marked in red, and the combinations for which the achieved value was lower than the target one are marked in blue. More intense colors indicate larger errors.

It could be noticed that – depending on the imbalance ratio – the optimization results for different criteria vary. In most of the tested cases, for highly imbalanced data, both in the case of the $F1$ and $N2$ criteria, it was not possible to achieve low values of the criterion (high difference values shown in red). On the other hand, for balanced problems, high criteria values were not achieved in the optimization task (low difference values visible in blue).

From the observations, it may be concluded that the difficulty measures are strongly related to the imbalance ratio of the problem. For a specific degree of imbalance, there is a very narrow range of values that the measure of difficulty can obtain. Hence, potentially there may be problems with bringing the criterion to a given value.

Classification Accuracy. Figure 2 shows the average quality of the classification expressed in the *balanced accuracy score* metric. The vertical axis of the heat map shows the successive values of the *imbalance ratio* parameter, while the horizontal axis shows the successive values of the *criterion target* parameter.

Let us notice that the imbalance ratio has a large impact on the classification quality. The lowest quality of the classification can be seen with the low share of

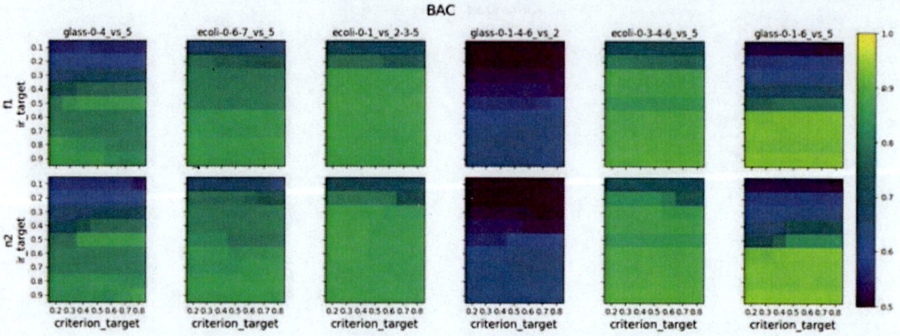

Fig. 2 Classification results after optimization procedure

the original minority class. For some of the datasets (e.g. `glass-0-4_vs_5`) the highest *BAC* metric values are obtained for balanced datasets, when *ir_target* = 0.5. For others, the highest classification quality is obtained for over-balancing the problem – for the cases when original minority class samples after oversampling states the 70–90% of output problem samples.

There are noticeable differences in the metric results for different values of the *criterion target* parameter – e.g., for the set `glass-0-4_vs_5` and *imbalance ratio* = 0.5, better results are achieved by sets in whose measures of complexity F1 and N2 were optimized to the value of 0.4–0.8. However, for the same dataset and imbalance ratio of 0.9, the sets achieved better results after resampling to the F1 measure close to zero and the N2 measure close to one. There is a more significant impact of the *imbalance ratio* parameter on the improvement of the result than the *criterion target* parameter.

3.2 Experiment 2: Dependency Research for Resampling Methods

Figure 3 shows the results for two selected oversampling algorithms – Random Oversampler and SMOTE, while the Fig. 4 shows the results for two selected undersampling methods – Random Undersampler and Cluster Centroids.

The graphs contain a summary of the classification results expressed in a *balanced accuracy score* (vertical axis) and the problem's complexity (horizontal axis), expressed in 22 measures. Each point in the graphs represents the difficulty value of one data set and the classification quality obtained after the data perturbation. The vertical line shows the standard deviation from the cross-validation of the evaluation loop.

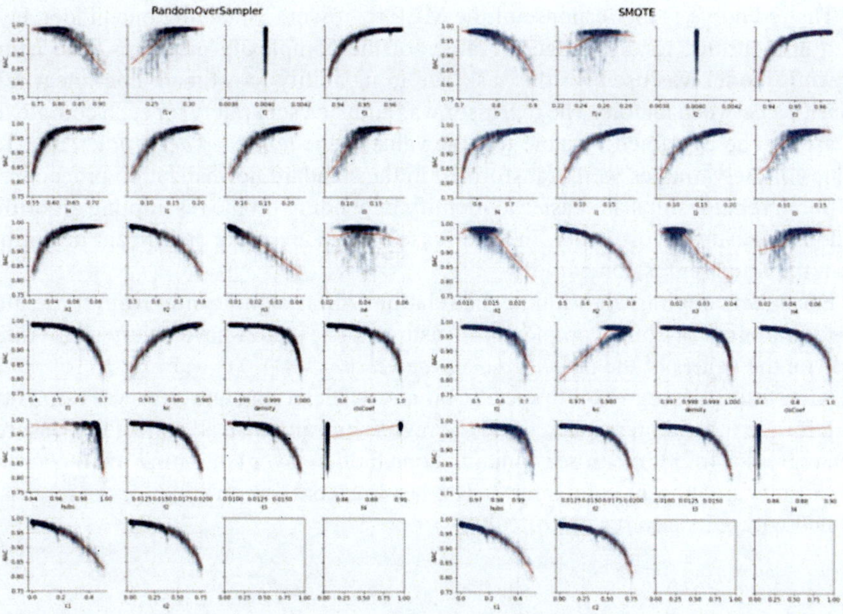

Fig. 3 Dependency in oversampling algorithms

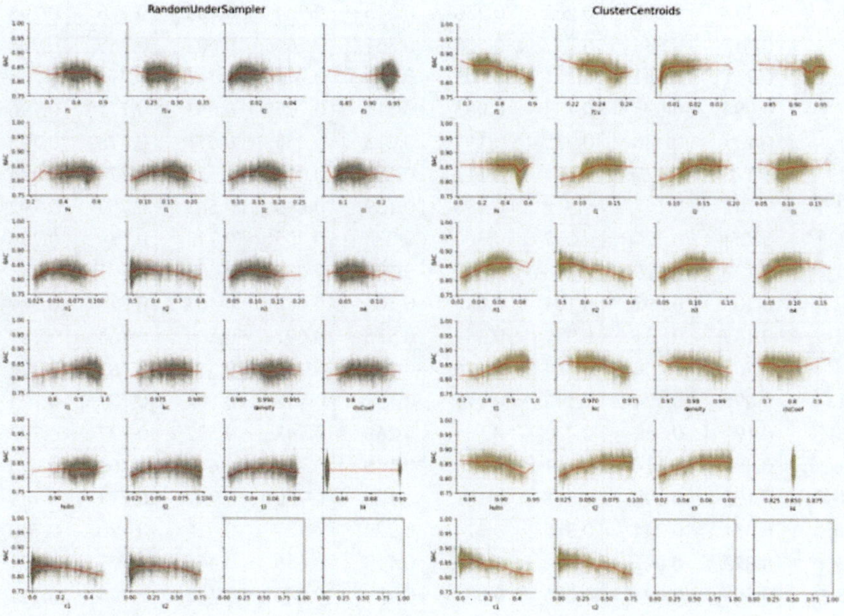

Fig. 4 Dependency in undersampling algorithms

The red curve is the response of the MLP regressor, containing one hidden layer with 30 neurons, for the range of values of the complexity measures. The neural network model was used for this task due to its ability to estimate non-linear relationships between factors. The regressor was adjusted separately for each complexity measure. The explained variable was the value of the *balanced accuracy score*. The values of the variables were transformed in the standard normalization procedure.

Some relationships are easier to identify depending on the resampling algorithm and the measure of difficulty. In other cases, there are more significant deviations from the fitted regression curve.

For undersampling algorithms, the relations differ significantly from those after oversampling. For some complexity measures, their values have a very slight influence on the values of the *balanced accuracy score*.

After fitting the regressor to specific points, its mean absolute error was estimated, both for the individual measures and the regressor trained on all collected measures. The regressor for all measures contained one hidden layer consisting of 50 neurons. The results are presented in Table 2. The last line contains an error for the regressor trained with all measures of difficulty.

Table 2 Mean Absolute Error of MLP regressor

measure	oversamplers					undersamplers			
	ros	smote	svms	bls	as	rus	cc	cnn	oss
F1	0.273	0.215	0.262	0.256	0.215	0.724	0.465	0.776	0.795
F1v	0.448	0.584	0.717	0.566	0.331	0.764	0.602	0.769	0.794
F2	0.711	0.681	0.381	0.705	0.670	0.741	0.541	0.778	0.796
F3	**0.094**	**0.095**	0.133	**0.083**	0.115	0.777	0.663	0.771	0.796
F4	0.137	0.186	0.235	0.197	0.215	0.755	0.671	0.778	0.795
L1	0.126	0.127	0.174	0.113	0.132	0.686	0.526	0.771	0.794
L2	0.142	0.147	0.218	0.130	0.139	0.695	0.533	0.769	0.792
L3	0.254	0.327	0.577	0.300	0.233	0.765	0.773	0.779	0.797
N1	0.132	0.419	0.258	0.311	0.591	0.683	0.494	0.771	0.788
N2	0.103	**0.092**	0.101	**0.083**	0.104	0.686	0.463	0.776	0.789
N3	0.137	0.431	0.279	0.344	0.575	0.697	0.504	0.764	0.789
N4	0.705	0.406	0.654	0.426	0.296	0.766	0.586	0.781	0.796
T1	**0.095**	0.174	0.172	0.175	0.197	0.712	0.504	0.777	0.795
LSC	0.106	0.351	0.222	0.184	0.260	0.748	0.522	0.777	0.792
Density	**0.097**	0.114	0.141	0.119	0.130	0.765	0.640	0.769	0.795
ClsCoef	0.105	0.107	0.108	0.101	0.128	0.760	0.757	0.777	0.795
Hubs	0.331	0.234	0.285	0.214	0.224	0.775	0.626	0.770	0.790
T2	**0.095**	**0.091**	**0.094**	**0.084**	0.110	0.686	0.464	0.777	0.794
T3	0.108	**0.094**	**0.096**	**0.089**	0.137	0.686	0.463	0.775	0.791
T4	0.460	0.684	0.722	0.693	0.666	0.785	0.785	0.784	0.799
C1	**0.095**	**0.091**	**0.097**	**0.083**	0.108	0.685	0.461	0.777	0.790
C2	**0.094**	**0.089**	0.102	**0.084**	0.118	0.686	0.462	0.777	0.793
All	**0.094**	**0.090**	**0.087**	**0.082**	0.112	0.622	0.443	0.770	0.798

The table emphasizes the results for which the MAE error was less than 0.1. Among the resamplers for which the estimation error is minor, there are four of the studied oversampling methods. The lowest error was observed when all complexity measures were used. It can therefore be expected that it is possible to approximate the classification quality based on the problem's complexity. As noted in the first experiment, however, this will not be a simple relationship, where changing one measure will directly impact the quality of the classification.

4 Conclusion

Analyzing insight available in the results of the conducted experimental evaluation, one may find evidence indicating the low legitimacy of using problem complexity measures as proxy classifiers. It is true that there are highly correlated factors in this respect. However, this correlation does not allow for an unambiguous analysis transfer from the problem complexity space to quality space.

The first experiment allowed one to observe that the optimization of a single measure of difficulty during the oversampling procedure has no easily predictable effect on the classification quality, despite the earlier demonstration of the relationship between the quality of the classification and the optimized measures of difficulty.

In the second experiment, the relationship between the quality of the classification and the difficulty measures on a broader spectrum was examined – for 1000 resampling procedures performed on the same set. A close-functional relationship between the values of the measures under study was observed, particularly for the oversampling algorithms. The mapping of all metrics to the quality of the classification was also examined. The mutual predictive relationship between recognition quality and complexity metrics – according to the performed study – seems to be not just a simple, one-dimensional weighted proportion of input factors, and it is worth verifying it in further research with broader dependency modeling by applying other ways of data resampling.

Acknowledgement This work was supported by the Polish National Science Centre under the grant No. 2019/35/B/ST6/04442.

References

1. Barella, V.H., Garcia, L.P., de Souto, M.C., Lorena, A.C., de Carvalho, A.C.: Assessing the data complexity of imbalanced datasets. Inf. Sci. **553**, 83–109 (2021)
2. Bartz, E., Zaefferer, M., Mersmann, O., Bartz-Beielstein, T.: Experimental investigation and evaluation of model-based hyperparameter optimization. arXiv preprint arXiv:2107.08761 (2021)

Image Processing

Deep Learning for Pit Detection in Cherries

Piotr Garbat and Maciej Smyl

Abstract The following paper presents the results of studies on deep-learning methods for pit detection in cherries. The general concept of a vision system for cherry pits detection has been proposed. The two types of light were considered and characterized. The paper overviews popular CNN architectures and discusses how additional polarizing camera information can be integrated to improve prediction accuracy. Therefore, several early and late methods are presented and described. The results of cherry classification confirm the effectiveness of the proposed solution and allow us to indicate a better solution.

1 Introduction

Monitoring the food processing chain at different steps is a critical task. Recently, food products have been inspected and sorted using cameras for different spectral ranges: RGB, ultraviolet (UV), infrared (IR), and XRay [1]. This recent development allows for addressing many problems that can no longer be solved using sensor technology. For example, we can check fruits while harvested, and the same applies when frozen fruits are sorted in production [2, 3]. In cherries processing, many features, such as pit presence, decay, and insect damage, can impact the quality of these products [4, 5]. The effective sorting procedure could determine the quality of the final product. Monitoring the pit removal process minimizes the risk of unwanted fruit pits. In industrial applications, the sorter machines may remove the cherries with the whole or damaged pit from the product stream. Modern methods, including computed tomography (CT) [6], X-Ray imaging [7], or near-infrared spectroscopy [8], are being tested for detecting external and internal structures within fruit tissue. Currently available, the most reliable industrial systems are based on the X-Ray imaging method. This type of imaging is expensive because of the specialized equipment.

P. Garbat (✉) · M. Smyl
Instituite of Microelectronics and Optoelectronics, Warsaw University of Technology, Nowowiejska 15/19, St., Warsaw, Poland
e-mail: piotr.garbat@pw.edu.pl

R. Burduk et al. (eds.), *Progress on Pattern Classification, Image Processing and Communications*, Lecture Notes in Networks and Systems 766,
https://doi.org/10.1007/978-3-031-41630-9_9

A more convenient, faster, less complicated and more economical x-ray inspection system proposed by Haff et al. [7] uses a line-scan configuration array of photodiode-based X-Ray detectors. This solution allows classifying pitted cherries with rates of 97.3% with a total error rate of 3.5%.

An up-and-coming solution was proposed by Liang Pei et al. [8]. Near-infrared spectroscopy in the range 800–2600 nm was used as a fast and effective method for detecting pits and pit fragments in fresh cherries. The best result - 96% overall accuracy- authors achieved under the binary model using only four features at specific near-infrared wavelengths in reflection mode.

In past years, hyperspectral imaging (HSI) systems have been successfully demonstrated to be the right solution for inspecting quality parameters in food products [9]. The research conducted by Siedliska et al. [10] shows the possibility of using these techniques for the automatic optical sorting of cherry fruits, possibly separating full-value and cherry with the pit. In this solution, the hyperspectral transmission images of cherries were acquired in the visible and near-infrared (VNIR) range (450–1000 nm), and machine learning algorithms were used to classify them. The best model allows achieving 94.6% of correctly classified instances for fresh samples and 83.3% for frozen cherries. These results showed that hyperspectral imaging in transmittance mode could be an alternative to systems using ionizing X-ray radiation in online sorting systems.

The study is part of the project aiming to implement an industrial system for the automatic optical sorting of cherry fruits, possibly separating full-value and cherry with the pit. We evaluate solutions to simplify image acquisition techniques. That allows for a significant decrease in the cost of the industrial system. The proposed solution used imaging in transmittance and reflectance mode for VIS and NIR (850nm) range, polarisation imaging techniques, and deep learning methods or multimodal classification.

2 Image Acquisition System

This study investigates the concatenation of two approaches to implementing the camera-based acquisition system. The first allows imaging in transmittance in the NIR range, and the second in reflectance mode for visible spectrum with additional polarisation cues analysis. Data acquisition for the research was based on an industrial CMOS camera -Basler for extended VIS range with bandpass filter for NIR range 740–850 nm and LED light source 835 nm (Fig. 1) and polarization camera with the light source in VIS range. The cherries are placed on a hole matrix. The shape of the holes is designed for the most efficient light propagation in fruits.

In transmission mode, the directional configuration of light was concerned. In this case, LEDs are located in axes of symmetry of holes in the matrix with cherries. Each hole is illuminated by exactly one LED placed directly under it. As a source 1W 835 nm LED was used.

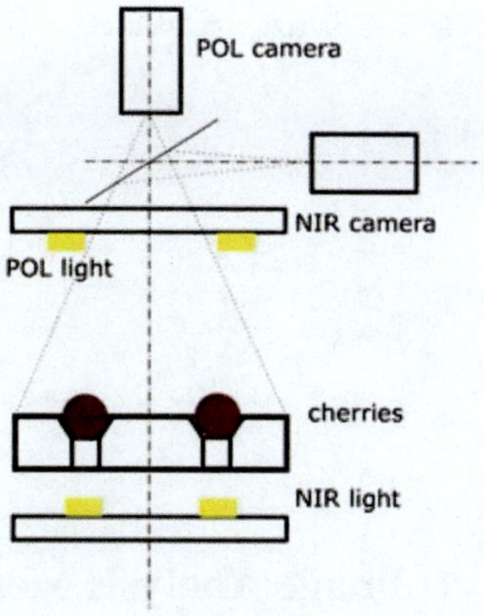

Fig. 1 General concept of a vision system for cherry pit detection

In the second path with polarization analysis, cherries are lighted with an area of white light emitters (spectral range 450–835 nm) from the camera direction.

The propagated and reflected light is directed to the specific cameras using a 50/50 beam-splitting device.

3 Datasets

The developed image acquisition setup allowed the preparation of data sets necessary for developing classification methods. The data sets allow us to train classification models and verify their effectiveness of them. The sets consist of fruit images in two classes - with and without pits gathered for both imaging modes. Each image consists of two images: NIR - image captured with Basler Ace camera, POL - polarization image captured with Lucid Phoenix Polarization camera.

The data set includes:

1. 650 images of pitted objects,
2. 1005 images of objects with pit.

The main purpose of the research was to use information from both cameras before starting the process of teaching the neural networks. All images are normalized and resized to the same dimension. The data set was prepared at the Institute of Microelectronics and Optoelectronics at the Faculty of Electronics and Information Technology.

Sample IR pictures prepared for training and the corresponding VIS-POL from class "0" and "1" are presented in Fig. 2.

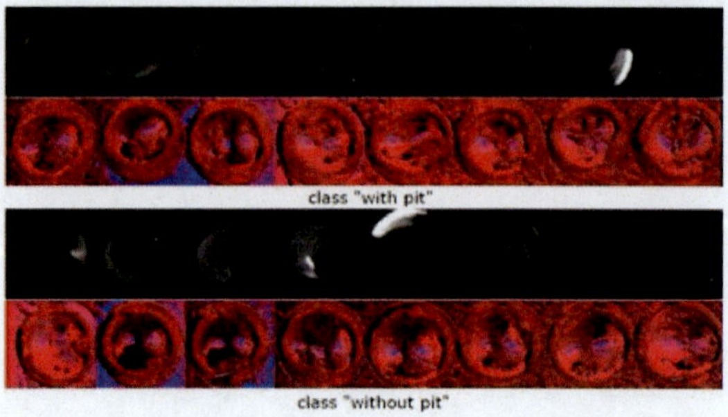

Fig. 2 Polarization Dataset.

4 Image Analysis Method

The main purpose of the research was to use information from both imaging modes: transmission and reflectance.

In recent years, many methods have been developed to increase the accuracy of prediction by combining predictions from multiple models trained separately for the same input (Ensemble Learning) [12], or combining predictions from multiple observations of the same scene [11].

In this study, we implemented methods that were based on the combination of predictions from infrared and polarized images. The methods can be divided into three main strategies:

- early fusion strategy,
- strategy of averaging the distribution of probabilities around certain classes,
- features merging strategy,

The commonly used ResNet50 architecture [13] was chosen as the base model. The learning was performed using the pre-trained for the ImageNet dataset.

The initial training of the models (pre-trained) was carried out separately for images from the IR camera (transmission mode) and the VIS-POL camera (reflection mode). The two base models that were received were used to combine the information obtained for both modalities. The ResNet50 model was trained for 100 epochs or until the loss no longer decreased. The stochastic gradient descent (SGD) optimizer was used with an initial learning rate of $lr = 0.005$, momentum $m = 0.8$, and weight decay $L2 = 0$. A batch size of $bs = 16$ images was used. In the training process, to artificially increase the volume of training data, the following operations were performed with a probability of 50%: random trimming, horizontal and vertical reflection, and change contrast, brightness, saturation, and hue.

The first strategy allows learning the model for all available data from different views or modalities. Most of the developed solutions in image classification use color photos consisting of three channels - red, green, and blue. It is possible to define an additional channel - carrying polarization information - and connect it to the input image.

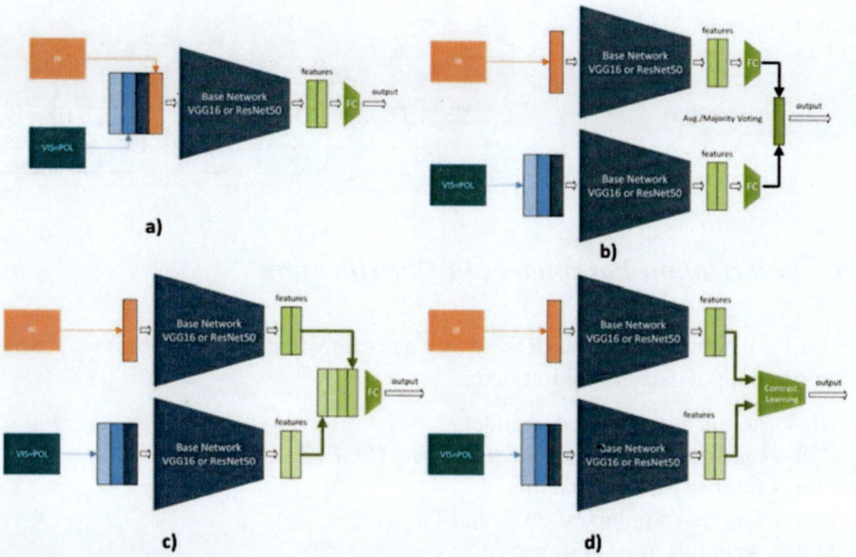

Fig. 3 Implemented strategies: a) merging the IR image and VIS-POL channels, b) averaging the prediction probabilities, c) a dual stream witch features concatenation architecture trained sequentially (for features extraction) and together (for merged prediction), d) a dual stream witch features concatenation architecture trained sequentially (for features extraction) and separately (for merged prediction)

The second strategy is widely used for aggregating multiple models trained for the same problem [12]. If a high correlation does not characterize the models, their averaging may increase the certainty of the classification. Such a strategy is often called late fusion [11].

In the third strategy, the features of two representations from base models are combined into one tensor and then used for classification by an additional neural network (fully connected layers). The diagram of the architectures created and used in the next chapter is shown in Fig. 3.

5 Experiments

The experiments tested the effectiveness of various methods of combining information from IR and POL cameras. The tests were performed using the ResNet50 network. The diagram of the data flow in the experiment is shown in Fig. 4.

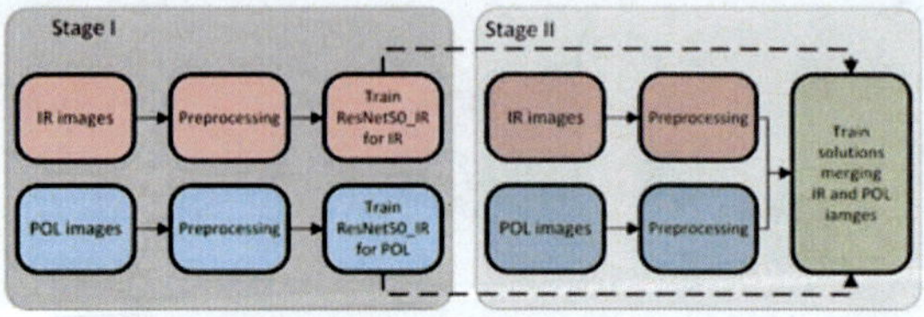

Fig. 4 Diagram of the flow of data in the experiment.

5.1 Polarization Parameters in Classification

For the base network solution (ResNet50), a learning and testing process was carried out for five input data configurations:

1. IR - baseline near-infrared channel,
2. POL - all three channels of the polarizing (POL) image,
3. POL - Intensity from POL image,
4. POL - Degree of polarization from POL image
5. POL - Azimuth from POL image,

Models based on the dataset were trained for both neural networks. The best prediction accuracy results were obtained for imaging in transmission for both networks with a slight advantage of ResNet architecture - accuracy equal 93% and 94%. In the case of using reflection imaging with polarization analysis, much worse results were obtained with accuracy equal to 89% and 86%. As expected, the best results were obtained using complete polarization information for this case.

5.2 Combining Information for Better Prediction

In this part, four methods were tested to combine information from two types of imaging. The learning performed on the IR image data sets was used as a reference. However, triple-multiplied single channels of polarization images were also used to ensure the possibility of using the weights of the model previously trained on the ImageNet set. Only ResNet50 has trained on IR and POL image data sets for this experiment. For this architecture, we study the following solutions:

- Method 1 is combining the IR image and POL channels (by adding to a single IR channel only the polarization azimuth layer and all POL layers),
- Method 2 is averaging the prediction probabilities,
- Method 3 is a dual stream witch features concatenation architecture trained sequentially (for features extraction) and together (for merged prediction),
- Method 4 is a dual stream witch features concatenation architecture trained sequentially (for features extraction) and separately (for merged prediction)

Methods 2, 3, and 4 used all polarization image channels examined and only the channel responsible for the azimuth of polarization. The results of the models trained

Table 1 Summary of results obtained by training the IR and POL sets separately and results of all developed methods "Method 1..4"

Dataset				
	Acc.	Sens.	Spec.	AUC
IR	0.94	0.97	0.88	0.97
POL	0.86	0.92	0.84	0.94
IR+POL	0.90	0.91	0.87	0.94
Method 1				
2 layers	0.94	0.96	0.90	0.98
4 layers	0.94	0.94	0.94	0.99
Method 2				
All POL channels	0.96	0.97	0.94	0.99
Only Angle of POL	0.96	0.95	0.95	0.99
Method 3				
All POL channels	0.94	0.96	0.90	0.98
Only Angle of POL	0.94	0.96	0.90	0.98
Method 4				
All POL channels	0.96	0.99	0.90	0.99
Only Angle of POL	0.95	0.95	0.95	0.99

based on the total three channels contained information on the intensity, degree, and azimuth of polarization. The comparison of the obtained results is presented in Table 1.

6 Summary

This study proposed a method for classifying objects between two classes (with the pit and the pit removed) based on multimodal data. Objects are classified based on images made with an infrared and polarizing camera. Using only the information from the NIR images, the accuracy is approximately 0.94, which indicates a very high classification result. Using only POL photos gives an accuracy equal to 0.84. In the case of polarization images, the channel carrying information about the polarization azimuth turned out to be the most advantageous in terms of prediction. The feature fusion architectures worked best, whereby both data sets were trained separately using ResNet50, and then the combined feature maps were used to train the fully connected layer itself (Acc 0.96). A similarly high result was obtained by averaging the probability of distribution between classes for both data sets (IR and POL).

Acknowledgement This work was supported by the National Centre for Research and Development, project POIR.01.01.01-00-1045/17.

References

1. Bhargava, A., Bansal, A.: Fruits and vegetables quality evaluation using computer vision: a review. J. King Saud Univ. Comput. Inf. Sci. **33**, 243–257 (2018)
2. Hameed, K., Chai, D., Rassau, A.: A comprehensive review of fruit and vegetable classification techniques. Image Vision Comput. **80**, 24–44 (2018)
3. Gongal, A.A., Karkee, S., Manoj Zhang, Q., Lewis, K.: Sensors and systems for fruit detection and localization: a review. Cmput. Electron. Agric. **116**, 8–19 (2015)
4. Qin, J., Lu, R.: Detection of pits in tart cherries by hyperspectral transmission imaging. Trans. ASAE **48**, 1963–1970 (2005)
5. Siedliska, A., Zubik, M., Baranowski, P., Mazurek, W.: Algorithms for detecting cherry pits on the basis of transmittance mode hyperspectral data. Int. Agrophys. **31**, 539–549 (2017)
6. Donis-González, I.R., Guyer, D.E., Kavdir, I., Shahriari, D., Pease, A.: Development and applicability of an agarose-based tart cherry phantom for computer tomography imaging. J. Food Meas. Character. **9**(3), 290–298 (2015)
7. Haff, R.P., Pearson, T.C., Jackson, E.: One dimensional linescan x-ray detection of pits in fresh cherries. Am. J. Agric. Sci. Technol. **1**, 18–26 (2013)
8. Pei, L., Roberto, M., Riccardo, M., Light, D., Haff, R.P.: Detection of pits and pit fragments in fresh cherries using near infrared spectroscopy. J. Near Infrared Spectrosc. **25**, 196–202 (2017)
9. Lu, Y., Huang, Y., Lu, R.: Innovative hyperspectral imaging-based techniques for quality evaluation of fruits and vegetables: a review. Appl. Sci. **7**(2), 189 (2017)
10. Siedliska, A., Baranowski, P., Zubik, M., Mazurek, W.: Detection of pits in fresh and frozen cherries using a hyperspectral system in transmittance mode. J. Food Eng. **215**, 61–71 (2017)
11. Roitberg, A., et al.: Analysis of deep fusion strategies for multi-modal gesture recognition (2019)
12. Hu, X., et al.: Combining models is more likely to give better predictions than single models. Phytopathology **105**, 1174–1182 (2015)
13. He, K., Zhang, X., Ren, S., Sun, J.: Deep residual learning for image recognition. In: Proceedings of the IEEE Conference on Computer Vision and Pattern Recognition, pp. 770–778 (2016)

Image Analysis System for Augmented Reality Games

Piotr Garbat, Grzegorz Galiński, Michał Żakowski, Mateusz Bieniek, and Maciej Lasocki

Abstract This article presents the idea of a vision system allowing the analysis of images from multiple cameras for the needs of augmented reality systems. The presented solution allows for the effective determination of the positions and movement paths of people moving in urban space and their activities. The article presents the key modules of multi-view image analysis, including the gamer detection and tracking (MOT) subsystem and the user activity detection subsystem. The discussed subsystems have been tested in order to determine the effectiveness of the proposed algorithms for determining the movement path and recognizing activity.

1 Introduction

Augmented Reality (AR) has become very popular in many applications, including human assistance in training, tourist guidance, and entertainment. AR games are gaining more and more popularity. Location-based AR games gather the most significant number of players in the AR game area. The combination of location features and augmented reality (AR) connects a game world to the real world. The first games of this type, such as Ingress and Ingress Prime, are based on the "capture flag" scenario. The most popular titles, such as Pokemon GO, are typical example of collection games. These scenarios perfectly match the features of games based on geo-location and do not require an exact location in space. The newest propositions, Pikmin Bloom, Jurassic World Alive, Witcher: Monster Slayer, and Orna: A Geo-

P. Garbat (✉) · M. Żakowski · M. Bieniek
Instituite of Microelectronics and Optoelectronics, Warsaw University of Technology, Nowowiejska 15/19, St., Warsaw, Poland
e-mail: piotr.garbat@pw.edu.pl

G. Galiński
Instituite of Radioelectronics and Multimedia Technology, Warsaw University of Technology, Nowowiejska 15/19, St., Warsaw, Poland

M. Lasocki
Faculty of Architecture, Warsaw University of Technology, Koszykowa 55, St., Warsaw, Poland

R. Burduk et al. (eds.), *Progress on Pattern Classification, Image Processing and Communications*, Lecture Notes in Networks and Systems 766,
https://doi.org/10.1007/978-3-031-41630-9_10

RPG, used boldly other scenarios, for example, role-playing (RPG), adventure, and tour-based strategy. The more complicated gameplays could require more accurate localization and additional information about user activity. The popularity of this technology trend is related to the development of new technologies for creating and displaying augmented content. The most significant device for this technology is an ordinary smartphone. Among many features of AR technology, the most significant is integrating multimedia content within the natural environment. To ensure the impression of unity of these two elements, it is necessary to ensure accurate and effective positioning of the smartphone display. This task has two main subtasks: a) global user localization in indoor or outdoor environments (geolocalisation), b) accurate display position and orientation estimation. This study focuses on user localization (geolocalisation). Many techniques allow for determining user position in the real world. The different techniques reveal their advantages in different applications. In outdoor games, the most popular solution is using the Global Positioning System (GPS). However, other techniques are still being developed. In indoor applications, other solutions could be considered. For this case, the most popular are wireless technologies, dead reckoning technologies, and computer vision-based technologies. This study focuses on the computer vision-based localization method that could be used in outdoor and indoor scenarios.

2 General Concept

The proposed solution is based on a vision system using external cameras placed in the space where the game is played. It is one of the components of the designed architecture. Its task is to enable tracking the player's movement, recognizing selected activities, and sending information about his location in real-time, making smooth gameplay possible. The game system, continuously receiving information about the player's location, allows tracking progress and introduces new multiplayer interaction opportunities. The described system consists of three basic elements:

- a set of cameras enabling observation from many directions and many places,
- a module for detecting and tracking the movement of many objects simultaneously MV-MOT (Multi-View Multiple Object Tracking),
- and an activity recognition module MV-VAR (Multi -View Video Action Recognition).

2.1 Detection and Tracking Module

The first module was based on the FairMOT method [1]. This solution uses an artificial neural network of the encoder-decoder type for image processing to detect

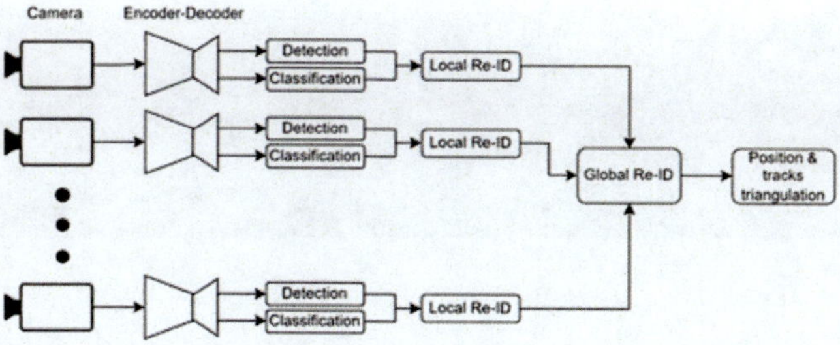

Fig. 1 MV-MOT

moving objects and their classification. As a result, it is possible to track the movement of any object and link its occurrences in different locations and moments. The designed architecture supports many cameras. They will allow one to observe a large area, which is the playing field. The goal of the detection system is to achieve the highest accuracy of tracking and recognizing the player, which will enable precise determination of his location on the game map. Not without significance here is the fact that there are many cameras, which entails the need to rebuild the architecture of the FairMOT algorithm so that it can link object instances between separate views. To handle the situation where there are many views of the same area, and there is a need to reidentify people between those views, a modular architecture was designed (Fig. 1). This architecture consists of detection and classification modules that are dedicated to certain cameras. This approach gives the possibility to adjust those algorithms to the specific perspective of the camera. Each module is equipped with a local Re-ID algorithm. The results of these algorithms are then propagated to the Global Re-ID module. This module is based on the Aligned Re-ID method [5]. The final result is a global ID that is assigned to certain objects in all of the views.

To determine the XYZ position of objects in the real 3D space, the internal parameters of the cameras have to be known, and the multi-camera system has to be calibrated to determine the external parameters (rotation and translations of the cameras relative to 3D space coordinates). To calibrate the camera system, a set of points is needed for which the actual XYZ position in 3D space is known, as well as their location in the images given by each camera. The calibration procedure must be performed separately for each location where the detection and tracking system will be operated. After calibrating the camera system, the XYZ position of the tracked objects can be calculated by the triangulation procedure, using the detection results in different video sequences (camera views). Figure 2 presents some example tracks detected by the described solution.

The developed solution has the capability to detect, track and locate pedestrians as well as other objects. During tests, several types of people and objects in an urban environment were considered for classification.

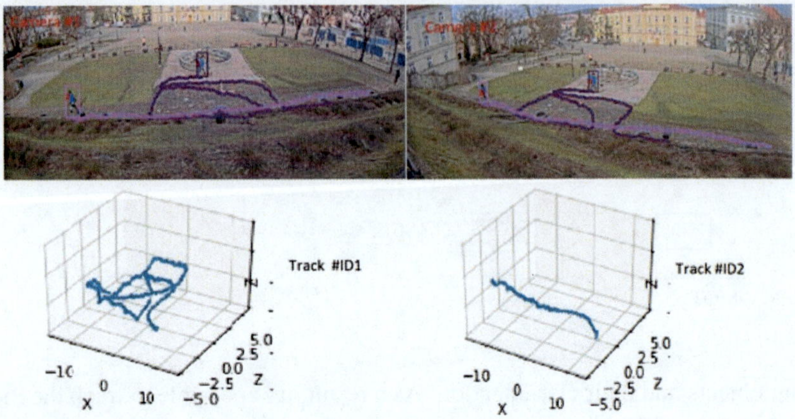

Fig. 2 Example tracks detected by MV-MOT solution

2.2 Global Re-ID

The Global Re-ID module of the discussed system is based on the Aligned Re-ID method [5]. In the first stage of this method, the deep convolutional neural network 'Resnet50' was adopted to extract a feature map from every image. Then global and local features were obtained from feature maps using global pooling and horizontal pooling, respectively. Finally, the distance between two-person images was calculated as the L2 distance of global and local features.

Local features are a representation of horizontal parts of a person's images. The distance between them is calculated dynamically so that the same fragments of the human body are compared with each other. The role of the global feature is to summarize the entire image. It is worth noting that local features are applied only at the learning stage. The inference stage is solely based on the global feature. As mentioned in [5], the global feature is more focused on a person's body than a picture background as a result of using local features in the learning stage. At the same time, the omission of local features at the inference stage is not associated with a significant decrease in the method's effectiveness.

It should be noted that two special cases can be observed during the operation of the image analysis system. (1) The person visible in one camera is not visible in the other, there is no match. (2) There are at least two similarly good matches of various persons between cameras. The Aligned Re-ID is able to determine the best match for every person's image, but it is not able to assess whether the match is correct. Thus, we proposed an approach to handle this issue. The solution is based on the following distance comparison metrics:

$$\frac{\left(\mathrm{dist}\left(q, g_{(2)}\right) - \mathrm{dist}\left(q, g_{(1)}\right)\right)}{\mathrm{dist}\left(q, g_{(1)}\right)} > \beta \tag{1}$$

$$\frac{\frac{1}{N}\sum_{i=1}^{N}\operatorname{dist}\left(q, g_{(i)}\right) - \operatorname{dist}\left(q, g_{(1)}\right)}{\operatorname{dist}\left(q, g_{(1)}\right)} > \beta \tag{2}$$

$$\frac{\frac{\operatorname{dist}(q, g_{(1)})}{\sum_{i=1}^{K}\operatorname{dist}(q_i, g^i_{(1)})} - \frac{1}{K}}{\frac{1}{K}} < \beta \tag{3}$$

where $\operatorname{dist}\left(q, g_{(i)}\right)$ is a distance between query image q and i-th best match $g_{(i)}$, N is a number of available $g_{(i)}$ images, K is a number of query images q_i, β is the acceptance threshold if exceeded, the match is considered invalid.

2.3 Action Recognition

Another system component is the player action recognition module [3]. During the game, the player is given a task to perform a specific action.

The set of actions to be detected consists of simple activities everyone can perform and don't require any prior knowledge. It includes arm movements ('one arm up', 'two arms up'), changing body position ('sitting down', 'crouching') or interacting with items ('taking a photo'). Additionally, the system is able to detect if an action has been performed or no action has been performed at all. It allows the game engine to assign tasks to the players. By performing actions in specified locations the players gain points and progress through the game scenario.

Correct execution and recognition of the action will allow the system to assign the player to the appropriate character in the character tracking system. One action can synchronize a maximum of one player per location.

The action recognition module tracks activities performed by all detected people in the scene. It works on the camera streams as well as outputs of the detection, tracking, and Global Re-ID modules described above. Data from each camera flow independently through the module up to a certain point at the end, where it is merged and final predictions are made. The outputs are action probabilities for each tracked person in the scene.

The module invokes the detection continuously, with a certain interval. It looks at 2.5 s long video sequences, where the end of the sequence is the current moment. The detection is performed for each person independently.

The first step of the pipeline is data preparation. This step cuts out an object of interest from video frames using bounding box data from the tracking module. The algorithm that has been implemented determines a minimum area covering all person bounding box positions for the 2.5 s long sequence. It will act as a mask. If the resulting rectangle is too narrow or wide, it's modified to be more square. Then, the mask is applied to the video sequences for each person and camera, followed by a center square crop.

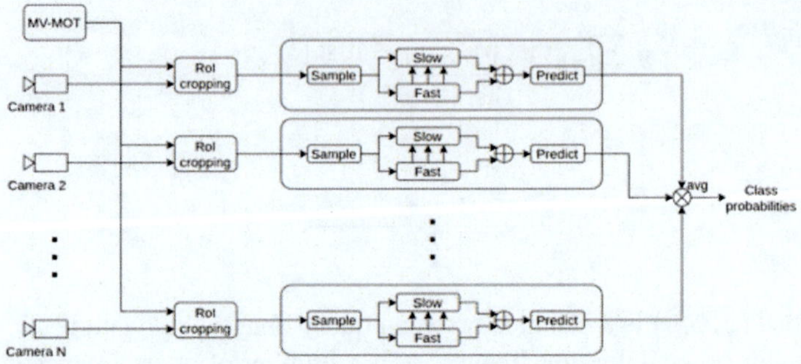

Fig. 3 Action recognition module

The next step is the actual action recognition (Fig. 3). The main ingredient is a deep convolutional neural network 'SlowFast' introduced in [4]. It is based on the idea that video sequences contain spatial and temporal information, and it can be separated to some extent to increase the accuracy and performance of the reasoning about actions.

The input data is sampled temporally with a different step for each component to meet the frequency requirement. The steps were chosen as 8 and 2 for the slow and fast components.

The outputs of the components are concatenated and fed through a fully connected layer. Finally, the Softmax layer produces a probability distribution over the action classes.

The module performs the two above steps for each camera. The last operation computes an average of the softmax probabilities over all cameras for each person. This gives the final distribution, and the class with the maximum probability is chosen.

3 Experiments

A methodology similar to that used in the Multiple Object Tracking (MOT) Challenge [2] is used to evaluate the system's performance. The basic metrics for evaluating the object detection quality in a single image are the number of correct detections (true positives, TP), the number of undetected objects (false negatives, FN), and the number of incorrect detections (false positives, FP).

For proper object tracking through subsequent frames of a video sequence, the given object must always have the same identifier assigned, meaning the system recognizes it as "the same object" in subsequent frames. To evaluate the system in terms of object tracking quality, the number of identity switches (IDSW) is also calculated in each video sequence frame, in addition to the TP, FN, and FP [2].

Table 1 Summary of results for sample sequence

Parameter	Unit	MOT	WUT2022
IDF1	[%]	80,17	89,68
MOTA	[%]	77,26	87,47
MED 3D_ACC	[cm]		27,06
AVG 3D_ACC	[cm]		38,11

For the overall evaluation of the quality of the detection system, single measure is used that consider the results obtained for individual frames. Namely, this is multiple object tracking accuracy (MOTA).

$$MOTA = 1 - \frac{\sum_t \{(FN_t + FP_t + IDSW_t\})}{\sum_t GT_t} \tag{4}$$

where: t is the frame number, GT is the number of ground-truth objects, and FN, FP, and IDSW as described above.

The above measures evaluate the quality of object detection and tracking in a single video sequence. One of the system's tasks is to track many players within the playing field in real-time. For this purpose, a dataset was prepared by the project team in the form of video sequences recorded on the market square in Przemyśl - WUT2022. The data in WUT2022 are anonymized, and the main actors are volunteers. These data and the publicly available MOT17 dataset were used to train the neural network model. The obtained results were verified using the public MOT16 dataset and videos recorded by the team.

The results are presented in Table 1.

To evaluate the system in terms of object positioning accuracy in 3D space, the Euclidean distance between the ground-truth object position and the position determined by the system is calculated. The distance is calculated for each object. Also, the average value of these distances is calculated. Table 1 provides the overall 3D location accuracy.

To evaluate the system in terms of action recognition, the accuracy and precision were calculated. A single evaluation sample comprised N 2.5 s long video sequences containing a person performing an action, where N was the number of cameras. The accuracy was calculated as the proportion of samples with correctly predicted actions among all evaluated samples.

The precision was calculated as macro-averaged precision over all classes. Firstly, the number of correct detections (true positives, TP_k) and the number of incorrect detections (false positives, FP_k) were calculated for each class. They were used to compute a class specific precision P_k and macro-averaged precision:

$$P_k = \frac{TP_k}{TP_k + FP_k}, \quad AP = \frac{1}{K}\sum_{i=1}^{K} P_i \tag{5}$$

Table 2 Summary of results for action detection

Camera setup	Accuracy	AP
One Camera	88,69%	88,99%
Two Cameras	89,95%	93,30%

The results are presented in Table 2.

4 Conclusions

The trained model is effective in urban conditions, which resulted in the possibility of using the tracking results in work on the second component of the designed architecture - the action recognition system. The results of the evaluation of the MV-MOT module confirm the high effectiveness in comparison with state-of-the-art solutions [6]. The accuracy of three-dimensional position determination below 40 cm allows using the system in the game environment for precise localization of players in the virtual game environment. The action recognition module allows recognition of six specific actions (defined by the game developer) with accuracy higher than 85% and precision higher than 88%. The system properties would realize more complicated game scenarios than are available currently.

Acknowledgement This work was supported by the National Centre for Research and Development, project POIR.01.02.00-00-0204/20.

References

1. Zhang, Y., Wang, C., Wang, X., Zeng, W., Liu, W.: FairMOT: on the fairness of detection and re-identification in multiple object tracking. Int. J. Comput. Vision **129**(11), 3069–3087 (2021)
2. Milan, A., Leal-Taixe, L., Reid, I., Roth, S., Schindler, K.: MOT16: a benchmark for multi-object tracking (2016)
3. Carreira, J., Zisserman, A.: Quo vadis, action recognition? a new model and the kinetics dataset (2018)
4. Feichtenhofer, C., Fan, H., Malik, J., He, K.: SlowFast networks for video recognition (2019)
5. Zhang, Hao X., et al.: AlignedReID: surpassing human-level performance in person re-identification (2018)
6. Benchmark MOT16 Results. https://motchallenge.net/results/MOT16/det=Private&orderBy=Rcll&orderStyle=ASC. Accessed 12 Apr 2022

Coregistration of Selected Sequences in MRI Imaging Based on Edge Analysis of Image Difference

Mikołaj Najda, Rafał Obuchowicz, and Adam Piórkowski

Abstract This paper discusses the challenges of maintaining consistent image sequences during patient examinations. One of the key issues that arises during MRI is the patient's inability to maintain a constant position, thus resulting in image shifts that impede image fusion and subsequent diagnostic quality. To address this, coregistration is employed, which involves aligning two or more images based on a reference image to minimize shifts between them. This is a crucial step in preparing data for analysis and can lead to previously overlooked images becoming important components of medical solutions.

In this study, a novel measurement of coregistration effectiveness is presented that includes normalization, difference, Top-hat transformation filtering, and global thresholding to achieve image similarity. Mean squared error and the structural similarity index measure are utilized as references against which the described method is compared. The outcomes of the measurements are presented and discussed. The results demonstrate the efficacy of the proposed metrics of coregistration effectiveness in facilitating a more accurate and comprehensive analysis of MRI data. Due to the achieved results, the main purpose is feasible.

Keywords Coregistration · MSE · SSIM · Top-hat · MRI

M. Najda (✉) · A. Piórkowski
Department of Biocybernetics and Biomedical Engineering, AGH University of Science and Technology, 30-059 Cracow, Poland
e-mail: najdamikolaj@protonmail.com

R. Obuchowicz
Department of Diagnostic Imaging, Jagiellonian University Medical College, 31-501 Cracow, Poland

R. Burduk et al. (eds.), *Progress on Pattern Classification, Image Processing and Communications*, Lecture Notes in Networks and Systems 766,
https://doi.org/10.1007/978-3-031-41630-9_11

1 Introduction

Magnetic resonance imaging (MRI) allows imaging of all planes of the human body without the use of ionizing radiation. This non-invasive and harmless procedure makes it possible to assess the condition of the patient's internal organs and tissues and detect emerging tumors or changes that are indicative of their development. It is possible to carry out imaging in different sequences, which can result in data offering a more convenient approach to diagnostic problems. Due to the long acquisition time of MRI images, especially compared to other methods, the patient is not able to maintain a constant, unchanging position. This results in shifts in the obtained image series, which may limit the possibilities for further diagnosis, especially when performing image fusion.

Coregistration is an image processing procedure that aims to align images to a common coordinate system. The method of coregistration requires the extraction of image features, on the basis of which the algorithm will be able to estimate the correct geometric displacement to align the compared images. Correct execution of the matching process will allow the development of deep learning networks that can segment the area of interest or, more accurately, observe tissue pathology changes over time. Creating a correct image coregistration algorithm is a difficult task, but it is worthwhile in these times of digitalization of imaging methods in the field of medicine.

Surface methods exploit image surfaces, boundaries, and contours for feature extraction and quantification, utilizing either the Iterative Closest Point Algorithm [10] or the Correspondence Matching Algorithm [4]. *Correlation-based Methods* are applicable to coregistering images of the same modality and object as the extracted features are utilized to obtain cross-correlation coefficients for image coregistration [1]. Additionally, neural networks are gaining popularity as an effective coregistration technique. Artificial intelligence-based methods have been reported to be useful for the coregistration of unimodal and multimodal images [2]. This work proposes the use of a procedure which is able to measure the gradient between bone tissue visible on images obtained with different sequences. The procedure contains normalization, difference, Top-hat transformation [3], global thresholding, and counting of the extracted pixel values. The reference measures include mean squared error and the structural similarity index.

2 Materials and Methods

2.1 The Dataset

The dataset contains images that depict spinal, cerebral, and wrist imaging results (Table 1). For the purposes of the article, the visual steps for wrist images are shown. Images from the same series but acquired via different modalities were manually

Table 1 Image properties.

ID	Anatomy structure	Original matrix (px)	Modality
S	Spine	512×512	T1, T2
W	Wrist	512×512	T1, T2, STIR
H	Head	448×512	T1, T2, T2TIRM

aligned and then augmented using an integer translation within the range of [−5, 5] and a step size of 1. The purpose of this augmentation was to assess the efficacy of the alignment process, and no interpolation methods were applied.

2.2 Designed Procedure

Analysis of the color channel merging of the compared images obtained with T1 and T2 modalities (Fig. 1) led to the hypothesis that it was feasible to extract differences resulting from a lack of aligned bone tissue visible in the images. Differences between bone tissue in the images when a pair of T1 and T2 images are not overlapped are especially visible (Fig. 1). Subsequently, attempts to extract the region of interest were carried out using the ImageJ environment (version 1.53t National Institutes of Health, USA) [7].

The pair of 16-bit images were first normalized to 8-bit depth and subtracted using the absolute value to avoid conflicts in the correct sequence of subtraction. The difference of non-overlapping images revealed dissimilarity between bone tissues (Fig. 2).

The results of this difference were filtered using the Top-hat transform to emphasize the regions of interest, which in this case are the bone tissue differences. This method permits the radius value that is applied to the algorithm to be selected. The differences caused by different radius values are shown in (Fig. 3). A value of 4 was chosen since other values either hid or exposed too much of the region of interest.

The threshold method was selected by comparing manually prepared masks of the region of interest with the results of threshold algorithms (Fig. 4). The following threshold algorithms were applied: Otsu [5], Yen [8], and Intermodes [6].

Otsu's threshold method was chosen based on the highest Dice score (Table 2). The extracted features were then counted by iteration over a 2D array of pixels, thus leading to the final result.

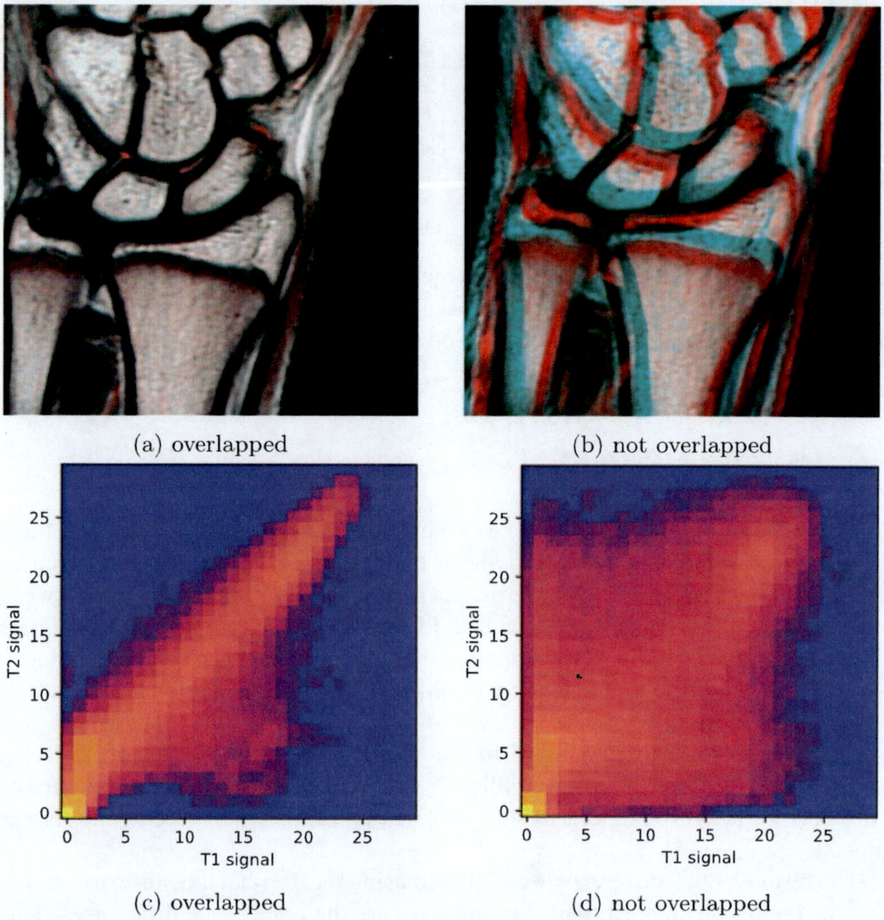

Fig. 1 Color channels (red, cyan) merged and signals comparison.

Table 2 Dice's coefficient results.

ID	Threshold method	Result
S	**Otsu**	**0.238**
	Yen	0.114
	Intermodes	0.223
W	**Otsu**	**0.747**
	Yen	0.647
	Intermodes	0.244
H	**Otsu**	**0.651**
	Yen	0.598
	Intermodes	0.634

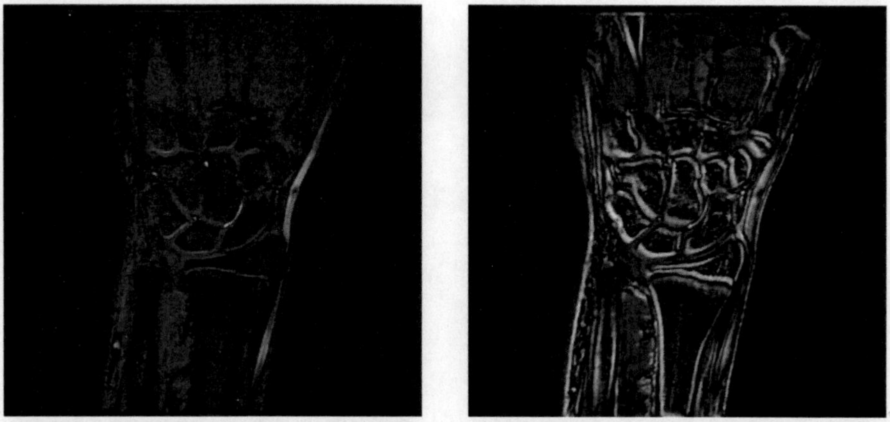

(a) |T1 - T2|

(b) |T1 - T2| augmented

Fig. 2 Difference of images.

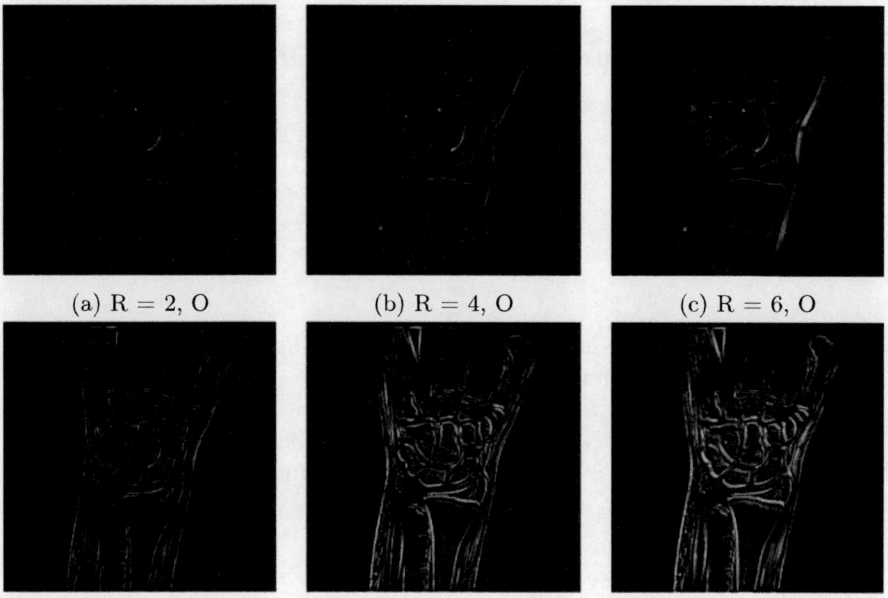

(a) R = 2, O

(b) R = 4, O

(c) R = 6, O

(d) R = 2, NO

(e) R = 4, NO

(f) R = 6, NO

Fig. 3 Comparison of the applied Top-hat transform with different radius values. R - radius, O - overlapped pair of images, NO - not overlapped pair of images.

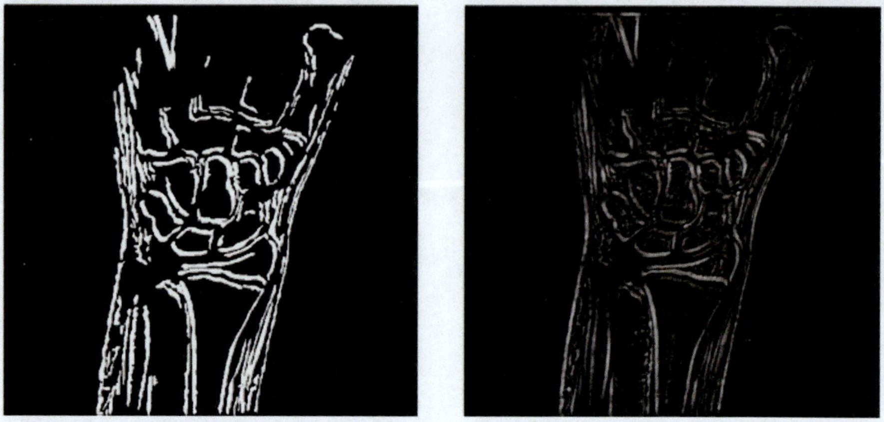

(a) mask (b) filtered image

Fig. 4 Masks and filtered images.

2.3 *Reference Measures of Image Alignment*

Mean squared error (MSE)is a method that allows for the measurement of the image-pair alignment error (Formula 1).

$$MSE = \frac{1}{m \cdot n} \sum_{i=1}^{m-1} \sum_{j=1}^{n-1} [(R(i, j) - M(i, j)]^2 \tag{1}$$

where: m - image height, n - image length, i, j pixel coordination points, R - reference image, M - image for alignment.

The structural similarity index measure (SSIM) [9] makes it possible to measure the similarity between images (Formula 2). If the compared images are obtained with different modalities, SSIM is a good choice for this approach as a reference measure for the designed procedure.

$$SSIM(x, y) = \frac{(2\mu_x\mu_y + C_1) + (2\sigma_{xy} + C_2)}{(\mu_x^2 + \mu_y^2 + C_1)(\sigma_x^2 + \sigma_y^2 + C_2)} \tag{2}$$

Where μ_x, μ_y are the mean values of compared signals, σ_x, σ_y, σ_x^2, σ_y^2 and σ_{xy} are the standard deviations, variances and covariance of respective signals, while C_1 and C_2 are small constant values to avoid the division by zero problem.

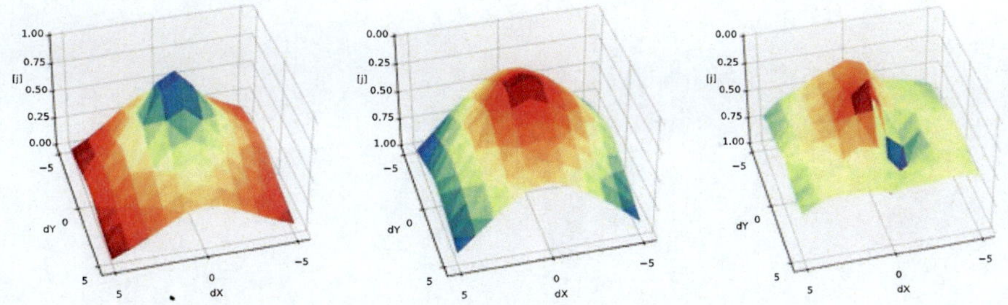

(a) S T2 SSIM (b) S T2 MSE (c) S T2 IP

Fig. 5 Results for Image S. IP stands for Implemented Procedure.

2.4 Results

The proposed method was implemented using ImageJ macros and was compared to SSIM and MSE to ensure its efficacy. The Otsu threshold method was selected after comparing manually prepared masks of the region of interest to the results of several threshold algorithms, including Otsu, Yen, and intermodes. The comparison was made using the Dice similarity coefficient. The extracted features were then counted by iterating over a 2D array of pixels. The normalized results are shown on 3D plots.

3 Summary

This study investigated the potential of using the identification of bone tissue differences in medical imaging. The implemented procedure involved normalizing image depths, subtracting images, applying a Top-hat transform filter, and using Otsu's threshold method to extract features. The performance of the implemented procedure was evaluated using two commonly used metrics, SSIM and MSE, as well as a custom procedure developed for this study. The proposed method resulted in significantly reduced regions in all examples, thus allowing for improved prediction of correct translations (Fig. 5 and Fig. 6). However, a slight deviation was observed in the comparison of images S, H, T1 and T2, where the right alignment was indicated by a metric close to 0, whereas it should be close to 1. This discrepancy may be attributed to the varying amounts of tissue present in the compared modalities. Further studies are necessary to evaluate the generalizability of the implemented procedure to other types of medical images and different image acquisition protocols.

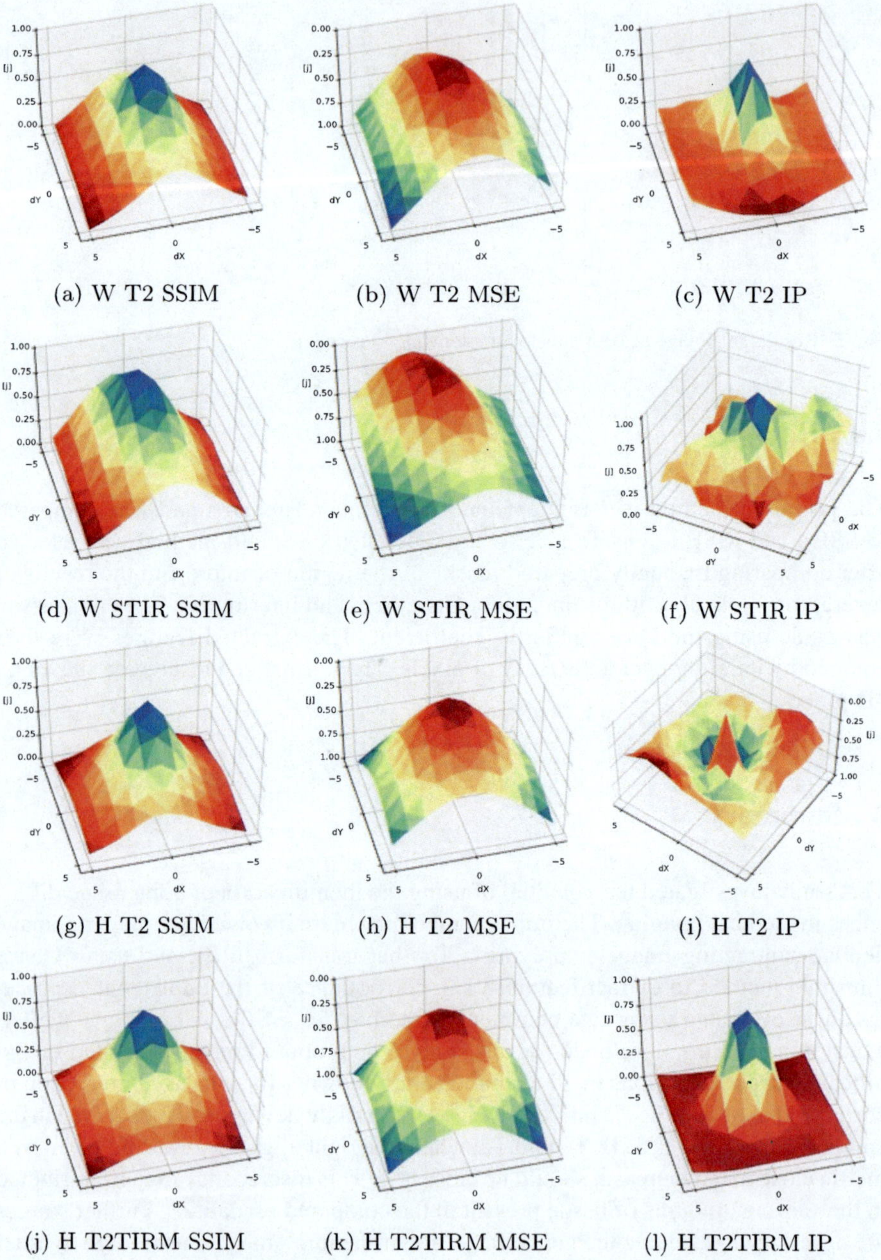

(a) W T2 SSIM (b) W T2 MSE (c) W T2 IP

(d) W STIR SSIM (e) W STIR MSE (f) W STIR IP

(g) H T2 SSIM (h) H T2 MSE (i) H T2 IP

(j) H T2TIRM SSIM (k) H T2TIRM MSE (l) H T2TIRM IP

Fig. 6 Results for Image W and Image H. IP stands for Implemented Procedure.

References

1. Bzowski, P., Borys, D., Guz, W., Obuchowicz, R., Piórkowski, A.: Evaluation of the MRI images matching using normalized mutual information method and preprocessing techniques. In: Choraś, M., Choraś, R.S. (eds.) Image Processing and Communications, pp. 92–100. Springer, Cham (2020)
2. Haskins, G., Kruger, U., Yan, P.: Deep learning in medical image registration: a survey. Mach. Vis. Appl. (2020). https://doi.org/10.1007/s00138-020-01060-x
3. Lange, H.: Real-time image stabilization based on a multiscale top hat filter. In: Dougherty, E.R., Astola, J.T. (eds.) Nonlinear Image Processing X, vol. 3646, pp. 305–318. International Society for Optics and Photonics, SPIE (1999). https://doi.org/10.1117/12.341097
4. Lv, J.J., Wen, G.J., Wang, J.Y.: A new feature matching algorithm for image registration based on feature similarity. In: 2008 Congress on Image and Signal Processing, vol. 4, pp. 421–425 (2008). https://doi.org/10.1109/CISP.2008.309
5. Otsu, N.: A threshold selection method from gray-level histograms. IEEE Trans. Syst. Man Cybern. **9**(1), 62–66 (1979). https://doi.org/10.1109/TSMC.1979.4310076
6. Prewitt, J.M.S., Mendelsohn, M.L.: The analysis of cell images. Ann. New York Acad. Sci. **128**, 1035–1053 (1966)
7. Schneider, C., Rasband, W., Eliceiri, K.: NIH image to ImageJ: 25 years of image analysis. Nat. Methods **9**, 671–675 (2012). https://doi.org/10.1038/nmeth.2089
8. Sezgin, M., Sankur, B.: Survey over image thresholding techniques and quantitative performance evaluation. J. Electron. Imaging **13**, 146–168 (2004). https://doi.org/10.1117/1.1631315
9. Wang, Z., Bovik, A., Sheikh, H., Simoncelli, E.: Image quality assessment: from error visibility to structural similarity. IEEE Trans. Image Process. **13**(4), 600–612 (2004). https://doi.org/10.1109/TIP.2003.819861
10. Zhang, C., Du, S., Liu, J., Xue, J.: Robust 3D point set registration using iterative closest point algorithm with bounded rotation angle. Signal Process. **120**, 777–788 (2016). https://doi.org/10.1016/j.sigpro.2015.01.021. https://www.sciencedirect.com/science/article/pii/S0165168415000547

Actinic Keratosis Prediction Based on Deep Learning Methods

Guan-Yi He, Chi-Ping Su, Chung-Shuo Chen, Yao-Sung Hsiang, Wei-Huan Hu, and Shin-Jye Lee

Abstract Actinic Keratosis (AK) is a type of skin lesion that typically appears on skin areas that are exposed to the sun. It is considered a precursor to squamous cell carcinoma (SCC), which is a common form of skin cancer worldwide. Early detection and treatment of AK are essential for effective management of SCC. Recent developments in deep learning (DL) have shown significant promise in improving the detection and diagnosis of AK and SCC. This study aimed to evaluate the use of YOLOv7, a state-of-the-art object detection model, in identifying AK lesions in skin images. We compared the accuracy and efficiency of YOLOv7 and YOLOv7-tiny models to determine the most suitable model for AK lesion detection. The results of the experiment were promising and showed that YOLOv7 can effectively identify AK lesions in skin images with high accuracy and speed. Furthermore, the study utilized the Grad-CAM technique to gain a deeper understanding of how the DL models detect AK lesions in skin images. We hope that this technology can assist dermatologists in making clinical decisions, leading to early treatment and prevention of AK, and ultimately preventing the development of skin cancer. Overall, the findings of this study highlight the potential of DL models in dermatology and their usefulness in improving clinical practice. With the increasing prevalence of skin cancer worldwide, the use of DL models may play a critical role in the early detection and prevention of skin lesions.

Keywords Actinic Keratosis prediction · Smart healthcare · Deep learning

---

G.-Y. He
Department of Dermatology, National Taiwan University Hospital Yunlin Branch, Yunlin, Taiwan

Graduate Institute of Clinical Medicine College of Medicine, National Taiwan University, Taipei, Taiwan

C.-P. Su · C.-S. Chen · Y.-S. Hsiang · S.-J. Lee (✉)
Institute of Management of Technology, National Yang Ming Chiao Tung University, Hinschu 300, Taiwan
e-mail: camhero@gmail.com

W.-H. Hu
College of Computer Science, National Yang Ming Chiao Tung University, Hsinchu 300, Taiwan

R. Burduk et al. (eds.), *Progress on Pattern Classification, Image Processing and Communications*, Lecture Notes in Networks and Systems 766,
https://doi.org/10.1007/978-3-031-41630-9_12

1 Introduction

AK is a type of skin lesion that commonly appears on sun-exposed skin areas and is considered a precursor to squamous cell carcinoma (SCC), which is a prevalent form of skin cancer worldwide. Skin cancer is a significant public health issue, with at least 40% of all cancer cases being related to skin cancer. In the United States, basal cell carcinoma (BCC) and SCC are the most common forms of skin cancer, with over 5 million new cases being diagnosed every year [1]. AK is especially prevalent in Australia, which has been reported to have the highest incidence of this lesion in the world [4]. The incidence of AK increases with age, and if left untreated, up to 10% of AK cases may progress to SCC. Therefore, early detection and intervention are crucial for effective management of AK and SCC.

Recent advances in artificial intelligence (AI) have shown promising results in improving the detection and diagnosis of AK and SCC. AI-based image analysis systems have demonstrated high accuracy and sensitivity in detecting AK and SCC [5, 6, 11, 16] from dermoscopic images. These systems can assist dermatologists in identifying lesions for biopsy and treatment, leading to earlier detection and improved patient outcomes. Additionally, AI can aid in the development of risk prediction models and personalized treatment plans for patients with AK, leading to more efficient and effective healthcare delivery. As such, AI has the potential to enhance the accuracy, efficiency, and cost-effectiveness of skin cancer management, thereby contributing to better health outcomes and improved quality of life for patients.

1.1 Deep Leaning for AK Classification

In recent years, AI models for classifying AK have been continuously studied. Among these AI models, deep learning (DL) approaches show outstanding performance on this task [11, 16].

In 2019, Nanni et al. experimented with different ensembles of support vector machines and convolutional neural networks (CNNs) for AK classification [11]. They introduced a data augmentation technique named FULL, and utilized the pre-trained ResNet101 model for training. Using FULL augmentation and ResNet101 model with fusion, the proposed model achieved an accuracy rate of 86.22%. Wang et al. also used CNN architecture for image recognition of AK and proposed a method called DL for AK (AK-DL) [16]. They trained the model on the HAM10000 public dataset and achieved an accuracy of 0.925 and an area under the receiver operating characteristic curve (AUC-ROC) of 0.887.

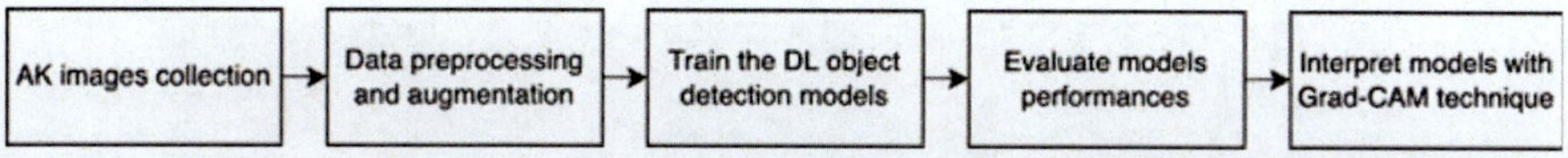

Fig. 1 Study framework

1.2 Deep Leaning for AK Detection

DL approaches have also been applied to detect AK using object detection techniques. With these techniques, the DL model not only can identify the presence of a particular type of lesion but also can locate and delineate it within the image.

Seung Han et al. proposed an object detection CNN architecture to identify and locate skin cancer, using the multiple clinical datasets [5]. The authors trained a R-CNN model with 1,106,886 images for skin cancer screening. Their proposed network model achieved an AUC score of 0.910. Hartanto et al. developed lightweight deep learning models using MobileNet v2 and Faster R-CNN to detect AK and melanoma on smartphones [6]. They trained the models using 300 images of AK and 300 images of melanoma. The experimental results demonstrated that the Faster R-CNN model outperformed MobileNet v2 with a test accuracy of 87.2%.

2 Methods

2.1 Data Collection

For this study, we retrospectively collected an image dataset of AK from the National Taiwan University Hospital. We selected 400 images that met the diagnostic criteria for AK, with resolutions between 1230×1504 and 3015×2505 pixels. Additionally, each image had a ground truth image containing the AK lesions region of interest (ROI), which was annotated by a dermatologist. The dataset included 432 ROIs. Two samples of AK images and their corresponding ground truth images are shown in Fig. 1. Finally, we randomly split the images in the dataset into 80% for model training and 20% for validation.

2.2 Data Preprocessing and Augmentation

Data preprocessing is essential to transform the data into a format that deep learning models can learn effectively. To ensure consistent input image size, each image was resized to resolution of 640*640 pixels. Additionally, standard normalization processing was performed on all channels of the images.

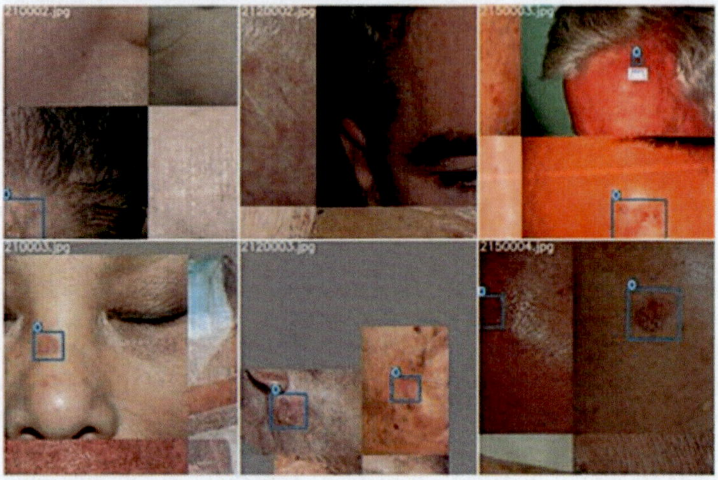

Fig. 2 Samples of augmented AK images

Since our dataset only contained 400 AK images, which is limited for training deep learning model, we implemented data augmentation techniques to improve the model's performance and avoid overfitting. The basic image augmentation techniques used in this study included randomly flipping images vertically and horizontally, scaling images randomly within 0.5 to 1.5 times, and applying random hue, saturation, and value augmentation.

In addition, a novel data augmentation technique named mosaic was applied. Mosaic augmentation combines four randomly selected training images into a single mosaic image, allowing the model to learn how to detect objects at a smaller scale than usual, which is ideal for small ROI in AK images. Another advantage of using this technique is that it enriches the background of the target AK lesions object, improving detection performance in AK images with complex backgrounds. Training on mosaic images also allows for better use of computational resources, as the GPU or CPU can process a larger batch of images at once. Samples of AK images after augmentation are shown in Fig. 2.

2.3 *Object Detection Model*

Medical imaging has become an important tool for the diagnosis and treatment of various diseases [10]. Object detection, a subfield of computer vision, plays a crucial role in medical image analysis by identifying and localizing important structures and anomalies within medical images [2, 7, 15]. In recent years, the You Only Look Once (YOLO) model has gained significant popularity as an efficient and accurate object detection model [3, 8, 14].

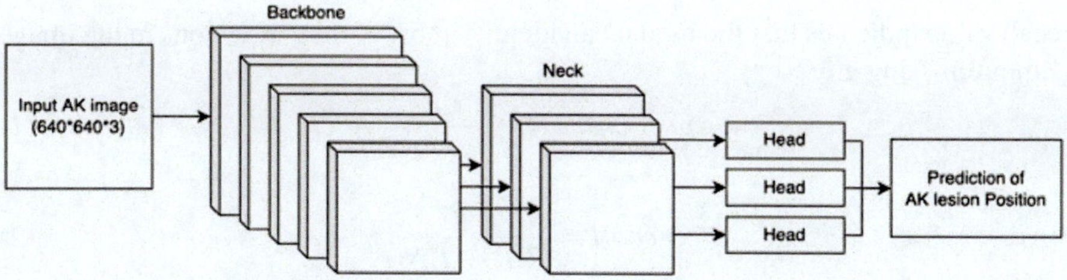

Fig. 3 Architecture of the YOLOv7 model.

YOLO (You Only Look Once) is a novel state-of-the-art object detection model that is well-suited for detecting AK lesions in skin images. Unlike traditional object detection models such as Faster R-CNN [12] that use multiple stages to detect objects, YOLO is a single-shot model that processes the entire image at once and outputs the detected objects and their corresponding bounding boxes. This makes YOLO faster and more accurate, which is particularly important in medical image analysis, where speed is often critical for timely diagnosis and treatment. In addition, YOLO can detect objects of varying sizes, which is crucial for detecting small AK regions in skin images. This is achieved through a feature pyramid network that extracts features at different scales, allowing the model to detect objects of different sizes in the same image [9]. Finally, YOLO has been extensively evaluated on various object detection benchmarks and has achieved state-of-the-art results in terms of accuracy and speed, making it a reliable choice for this study.

The purpose of this study was to develop an effective and precise automated approach for identifying AK lesions in skin images. To accomplish this objective, we assessed the latest versions of the YOLO model family, specifically YOLOv7 and YOLOv7-tiny [14]. We examined the trade-off between accuracy and efficiency by comparing these two models, with YOLOv7 serving as the baseline model and YOLOv7-tiny serving as the smaller and slightly less accurate model, but well-known for its high speed and low computational requirements. Through this comparison, we aimed to determine the most appropriate model for AK lesion detection in terms of both accuracy and speed (Fig. 3).

2.4 *Performance Metrics*

The YOLO models proposed in this study were evaluated based on several performance metrics, including precision, recall, and mean average precision (mAP).

Precision represents the ratio of correctly identified AK lesions to the total number of lesions detected by the model. A high precision value indicates that the model can accurately identify AK lesions without mistakenly detecting non-lesion regions as AK. Recall, on the other hand, measures the proportion of AK lesions correctly identified by the model among all the actual AK lesions present in the image. A high recall value indicates that the model can identify most of the AK lesions in the image without missing any.

$$precision = \frac{TP}{TP + FP} \tag{1}$$

$$recall = \frac{TP}{TP + FN} \tag{2}$$

where TP (True Positive) represents the number of cases where the model correctly detected an object with the detected bounding box has an intersection-over-union (IoU) overlap with the ground truth bounding box above a predefined threshold, FP (False Positive) represents the number of cases where the model incorrectly detected an object, and FN (False Negative) represents the number of cases where the model failed to detect an object that exists in the ground truth.

In addition to these metrics, we also calculated the mean average precision (mAP) to evaluate the models' performance across different levels of confidence thresholds. The mAP is a widely used performance metric for object detection models, and it calculates the average precision (AP) across different confidence thresholds. A high mAP indicates that the model can accurately detect AK lesions across a range of confidence thresholds.

$$AP = \frac{1}{11} \sum_{r} P_{interp(r)} \tag{3}$$

$$mAP = \frac{1}{C} \sum AP \tag{4}$$

where $P_{interp(r)}$ is the interpolated precision values and r is the recall value at each point on the PR curve, C is the total number of classes.

2.5 Model Interpretation

As an additional analysis to evaluate the performance of our proposed deep learning models for AK lesion detection, we incorporated the Grad-CAM technique [13]. This technique enables the generation of heatmap images that provide visual insights into how the models are making predictions. The heatmap images use a color gradient ranging from dark blue to yellow to dark red, which indicates the level of attention the model is giving to different areas of the image during the prediction process. Areas in the image with higher attention will be represented with warmer colors such as yellow and red, while areas with lower attention will be represented with cooler colors such as blue.

The Grad-CAM technique allows us to gain a deeper understanding of how our models are detecting AK lesions in skin images. By examining the heatmap images generated by the technique, we can identify which regions of the skin images are

Table 1 Performance comparison between two YOLO model architectures in AK detection.

Metrics Parameters	YOLOv7	YOLOv7-tiny
	36 million	6 million
Precision	0.93	0.95
Recall	0.80	0.73
mAP@0.5	0.87	0.85
FPS	238	370

FPS: frame per second

being given more importance by the model for accurate detection. Additionally, we can also identify any potential false positives or false negatives by analyzing the areas in the image where the model is focusing its attention.

3 Results

The performances of the YOLOv7 and YOLOv7-tiny model in AK detection are shown in Table 1. This table compares the performance of two different models, YOLOv7 and YOLOv7-tiny, in terms of their parameters, precision, recall, mAP at 0.5 IoU threshold, and frames per second (FPS). YOLOv7 has 36 million parameters, while YOLOv7-tiny has only 6 million parameters. The precision of YOLOv7 is 0.93, while YOLOv7-tiny has a higher precision of 0.95. However, YOLOv7 has a higher recall of 0.80 compared to 0.73 of YOLOv7-tiny. Both models have a similar mAP of 0.87 and 0.85. Finally, YOLOv7-tiny has a higher FPS of 370 compared to YOLOv7, which has an FPS of 238.

In this study, we utilized the Grad-CAM technique to generate heatmap images to interpret the model's predictions. Two images with AK lesion in the validation set were predicted by the trained YOLOv7 and YOLOv7-tiny model, generating four heatmap images shown in Fig. 4.

4 Discussion

The YOLOv7 and YOLOv7-tiny models performed well in AK lesion detection in this study. Typically, larger CNN architectures with more parameters tend to have better performance. However, in our experiment, the YOLOv7-tiny model, which has only one-sixth of the parameters compared to YOLOv7, achieved similar performance. This is mainly because AK lesion images are relatively simple compared to real-world images, which often contain multiple objects in complex backgrounds. As a result, CNN models can learn the features in AK images more easily.

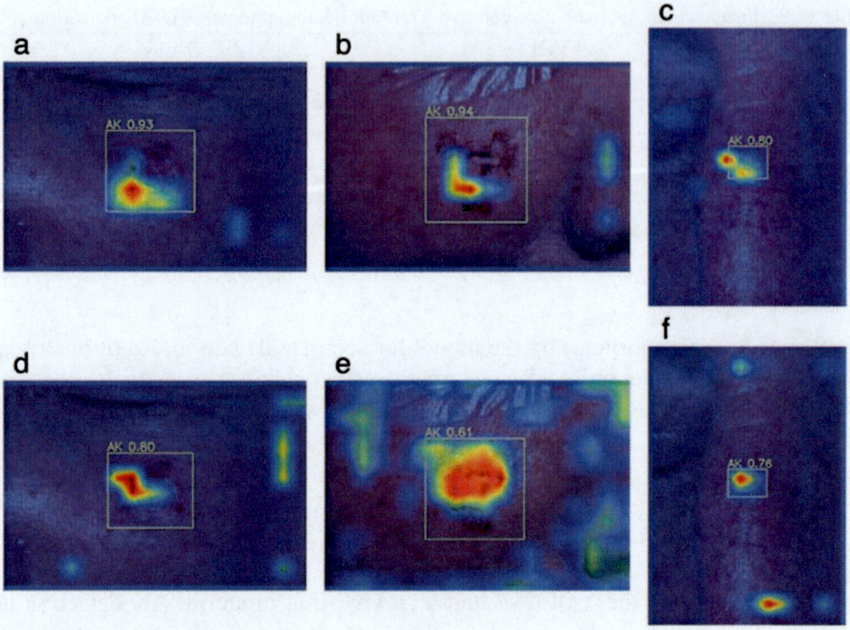

Fig. 4 Heatmap images generated through the Grad-CAM technique. Image a., b., and c. were obtained from the YOLOv7 model. Image d., e., and f. were obtained from the YOLOv7-tiny model.

The training time, processing time, and computational cost of the YOLOv7-tiny model are shorter, making it more efficient to use and easy to deploy on low-computational devices such as smartphones. However, YOLOv7 demonstrates better recall on this task, indicating that the model can correctly identify a higher percentage of all relevant instances in the dataset. In other words, it has a lower chance of missing a positive instance, which is usually a more critical metric for clinical usage. The selection between the two models ultimately depends on the specific needs and requirements of the AK lesion detection task at hand.

For future research on the application of DL in audiograms, we suggest that researchers first use a lightweight CNN model architecture and then gradually deepen the network according to performance to shorten the time spent on model training.

By incorporating the Grad-CAM technique into our evaluation process, we can improve the interpretability of our models and provide clinicians with greater confidence in their use for diagnosing and treating AK lesions. The insights gained from the heatmap images can also be used to refine the models and improve their performance further.

5 Conclusion

In this paper, we have demonstrated the feasibility and effectiveness of using the YOLOv7 model to detect AK lesions. Our proposed method achieved high precision, recall, and mAP scores of 0.93, 0.80, and 0.87, respectively. The model can also detect AK lesions in real-time with an impressive FPS of 238. Furthermore, our lightweight YOLOv7-tiny model can produce similar results with less training time, processing time, and computational cost, making it ideal for deployment on mobile or low computational devices. The heatmap images generated by the Grad-CAM technique revealed that both models classify AK images based on the lesion area, providing additional information to doctors and patients about how AI models infer AK images. This technology can assist doctors in making clinical decisions, and patients with AK can be treated and prevented early, helping to prevent the disease from developing into skin cancer. Overall, our findings highlight the potential of DL models in dermatology and their usefulness in improving clinical practice.

The main limitation of this study is our dataset. We used a dataset of AK images collected from a single source. Expanding the dataset to include images from multiple sources and incorporating different skin types and conditions could improve the robustness and generalizability of our model. Our future work will focus on validating our model on a larger dataset and in a clinical setting to assess its usefulness in diagnosing and treating AK lesions.

Acknowledgement This research was funded by Taiwan National Science and Technology Council (grant number MOST 111-2410-H-A49-019-) and National Taiwan University Hospital Yunlin Branch (NTUHYL 111.S031).

References

1. American Cancer Society. Key statistics for basal and squamous cell skin cancers. cancer.org (2021)
2. Al-Masni, M.A., Al-Antari, M.A., Park, J.M., Gi, G., Kim, T.Y., Rivera, P., Valarezo, E., Choi, M.T., Han, S.M., Kim, T.S.: Simultaneous detection and classification of breast masses in digital mammograms via a deep learning yolo-based cad system. Comput. Methods Programs Biomed. **157**, 85–94 (2018)
3. Bochkovskiy, A., Wang, C.Y., Liao, H.Y.M.: Yolov4: optimal speed and accuracy of object detection. arXiv preprint arXiv:2004.10934 (2020)
4. Chetty, P., Choi, F., Mitchell, T.: Primary care review of actinic keratosis and its therapeutic options: a global perspective. Dermatol. Therapy **5**(1), 19–35 (2015)
5. Han, S.S., Moon, I.J., Lim, W., Suh, I.S., Lee, S.Y., Na, J.I., Kim, S.H., Chang, S.E.: Keratinocytic skin cancer detection on the face using region-based convolutional neural network. JAMA Dermatol. **156**(1), 29–37 (2020)
6. Hartanto, C.A., Wibowo, A.: Development of mobile skin cancer detection using faster R-CNN and mobilenet V2 model, pp. 58–63 (2020)
7. Kaldera, H., Gunasekara, S.R., Dissanayake, M.B.: Brain tumor classification and segmentation using faster r-cnn. In: 2019 Advances in Science and Engineering Technology International Conferences (ASET), pp. 1–6. IEEE (2019)

8. Li, C., Li, L., Jiang, H., Weng, K., Geng, Y., Li, L., Ke, Z., Li, Q., Cheng, M., Nie, W., et al.: Yolov6: A single-stage object detection framework for industrial applications. arXiv preprint arXiv:2209.02976 (2022)
9. Lin, T.Y., Dollár, P., Girshick, R., He, K., Hariharan, B., Belongie, S.: Feature pyramid networks for object detection. In: Proceedings of the IEEE Conference on Computer Vision and Pattern Recognition, pp. 2117–2125 (2017)
10. Litjens, G., Kooi, T., Bejnordi, B.E., Setio, A.A.A., Ciompi, F., Ghafoorian, M., Van Der Laak, J.A., Van Ginneken, B., Sánchez, C.I.: A survey on deep learning in medical image analysis. Med. Image Anal. **42**, 60–88 (2017)
11. Nanni, L., Paci, M., Maguolo, G., Ghidoni, S.: Deep learning for actinic keratosis classification. AIMS Electron. Electr. Eng. (2019)
12. Ren, S., He, K., Girshick, R., Sun, J.: Faster R-CNN: towards real-time object detection with region proposal networks. In: Advances in Neural Information Processing Systems, vol. 28 (2015)
13. Selvaraju, R.R., Cogswell, M., Das, A., Vedantam, R., Parikh, D., Batra, D.: Grad-cam: visual explanations from deep networks via gradient-based localization. In: Proceedings of the IEEE International Conference on Computer Vision, pp. 618–626 (2017)
14. Wang, C.Y., Bochkovskiy, A., Liao, H.Y.M.: Yolov7: trainable bag-of-freebies sets new state-of-the-art for real-time object detectors. arXiv preprint arXiv:2207.02696 (2022)
15. Wang, H.J., Su, C.P., Lai, C.C., Chen, W.R., Chen, C., Ho, L.Y., Chu, W.C., Lien, C.Y.: Deep learning-based computer-aided diagnosis of rheumatoid arthritis with hand X-ray images conforming to modified total sharp/van der Heijde score. Biomedicines **10**(6), 1355 (2022)
16. Wang, L., Chen, A., Zhang, Y., Wang, X., Zhang, Y., Shen, Q., Xue, Y.: AK-DL: a shallow neural network model for diagnosing actinic keratosis with better performance than deep neural networks. Diagnostics **10**(4), 217 (2020)

A New Assessment of Convolutional Neural Networks for Texture Directionality Detection

Marcin Kociołek and Antonio Cardone

Abstract Image texture analysis is ubiquitous as it finds applications in many scientific fields of interest, including biomedical and material science. The detection of meaningful texture properties such as directionality remains a challenging task, due to the complexity of texture. We build upon our past work on the design of convolutional neural networks (CNNs) for texture directionality detection. The CNNs are trained on a library of synthetic textures with known directionality and varying perturbation levels. The present effort focuses on enhancing the training data through a new perturbation procedure and a more diverse set of synthetic textures. We study the performance of new CNN architectures, such as grouped CNNs, on the enhanced synthetic texture library. The results yield novel insight into CNN-based texture directionality detection. Shallow and grouped CNNs show better performance than deep CNNs, unlike previously. We discuss this performance shift and its implications, and suggest possible future work directions.

Keywords directionality detection · texture analysis · image processing

1 Introduction

Image texture analysis is challenging due to the complexity of texture local patterns, or textons [1]. Texture directionality is an important property whose study finds applications in several domains. In the past we proposed the use of convolutional neural networks (CNN) trained on synthetic textures for texture directionality detection [2]. The synthetic textures consist of patterns with known directionality and increasing

M. Kociołek (✉)
Lodz University of Technology Institute of Electronics Al., Politechniki 10, 93-590 Lodz, Poland
e-mail: marcin.kociolek@p.lodz.pl
URL: https://ie.p.lodz.pl

A. Cardone
National Institute of Technology, Software and Systems Division, 100 Bureau Drive, Gaithersburg, MD 20899, USA

R. Burduk et al. (eds.), *Progress on Pattern Classification, Image Processing and Communications*, Lecture Notes in Networks and Systems 766,
https://doi.org/10.1007/978-3-031-41630-9_13

Fig. 1 Instances of synthetic texture images from our synthetic texture library

levels of perturbations (i.e., blur and additive noise) to mimic real-life texture images (Fig. 1). The resulting library consists of textures whose directionality is discernable by a human observer.

The previously designed CNNs consisting of one convolutional layer (henceforth shallow CNNs) and of three convolutional layers (henceforth deep CNNs). The CNNs were trained, validated and tested on the synthetic texture library. Both shallow (SN1) and deep (DN2) CNNs were found to perform better than or comparably to state-of-the-art techniques. Qualitative assessment suggested that the CNNs trained on synthetic textures nicely generalize to real-life images [2].

In this paper we improve the training data by eliminating aliasing effects [2] and by using synthetic texture images of different sizes. Texture is a local property of images whose patterns typically have different sizes, and hence multi-size training textures more closely mimic real-life textures. Furthermore, we get additional insight by testing previously and newly designed CNNs on the enhanced training data, and by using additional activation functions [2]. Most of the new CNNs consist of grouped CNNs, in which the convolutions are performed independently for each group at any given layer. Grouped CNNs reduce the number of computations and enable a better representations of the data.

The paper is organized as follows. In Sect. 2, the related literature is surveyed. We describe the enhanced training data and the CNN architectures in Sect. 3. The results are presented in Sect. 4, and our conclusions and future work directions are discussed in Sect. 5.

2 Related Work

Texture directionality analysis plays a fundamental role in fields such as image retrieval and classification [3–5]. In the biomedical field, therapeutic approaches often rely on the characterization of texture directionality, as in placebo-controlled trials [6]. In material science and engineering, texture directionality is correlated to material properties [7]. In astronomy, high-throughput texture directionality analysis has been applied to solar images [8].

Texture directionality detection approaches rely on computational entities such as the Fourier transform [9] and the local gradient orientation (LGO) [10](implemented in the Fiji/ImageJ software [11]), the Radon transform [12], and the gray level co-occurrence matrix (GLCM) [13]. The design of the underlying models can be challenging, as the parameters need to be customized for the targeted data. Many models involve heavy computations and are not viable for high-throughput efforts.

Supervised machine learning can circumvent challenging model designs. We have tested CNNs trained on synthetic textures for directionality detection, and have identified shallow and deep CNNs with accuracy comparable to the state-of-the-art and less computational load [2]. Still, the design space of CNNs needs to be further explored.

Grouped CNNs are potentially effective for image analysis. Their convolutional layers are organized in groups with independently performed convolutions. They were introduced first in AlexNet [14], evidence shows improved classification accuracy [15]. One reason could be that grouped CNNs introduce a new dimension across the grouped convolutions, cardinality, with size proportional to the network accuracy [15].

A related topic is the equivalence of a cascade of smaller filters (e.g., smaller convolutions in a grouped CNN) to a larger filter (e.g., larger convolution in a shallow CNN). Given a large filter, an equivalent cascade of smaller filters has always fewer parameters, hence it can be found only under specific conditions [16]. As such equivalence is possible, it is worth studying grouped CNNs in this context.

3 Materials and Methods

The work presented in this paper is two-fold: we improve the training data used in our previous effort [2] and we test previously and newly designed CNN architectures on the enhanced training data, with additional activation functions.

3.1 Enhanced Training Data

The quality of training data is of paramount importance for the training of CNNs. In this work, we build an enhanced version of the synthetic texture image library.

First, we address the aliasing effects caused by averaging filters [2] by employing Gaussian filters instead [17]. Different blurring levels are obtained by varying the standard deviation σ within a range. The range was selected empirically by visual inspection by ensuring no aliasing effects are present at any given bar period and thickness combination. Table 1 shows the corresponding maximum Gaussian standard deviation σ values.

The rationale behind Gaussian filters is that they act as low-pass filters, as the Gaussian function yields a smoother impulse response. Instead, averaging filters cause sharp transitions and they act as step filters. Gaussian filters mitigate the spatial anisotropy associated with the filter shape, in our case a square, since the pixels away

Table 1 Bar periods, bar sizes and maximum σ for the Gaussian blur used in this paper

bar period	4	4	4	6	6	6	6	6	8	8	8	12	12	12
bar size	1	2	3	1	2	3	4	5	2	4	6	2	4	8
blur maximum σ	0	0	0	0	1	1	1	0	2	2	2	4	4	4
bar period	12	16	16	16	20	20	20	24	24	24	30	30	30	30
bar size	10	4	8	12	5	10	15	6	12	18	6	12	18	24
blur maximum σ	4	5	5	5	6	6	6	6	7	7	8	8	8	8

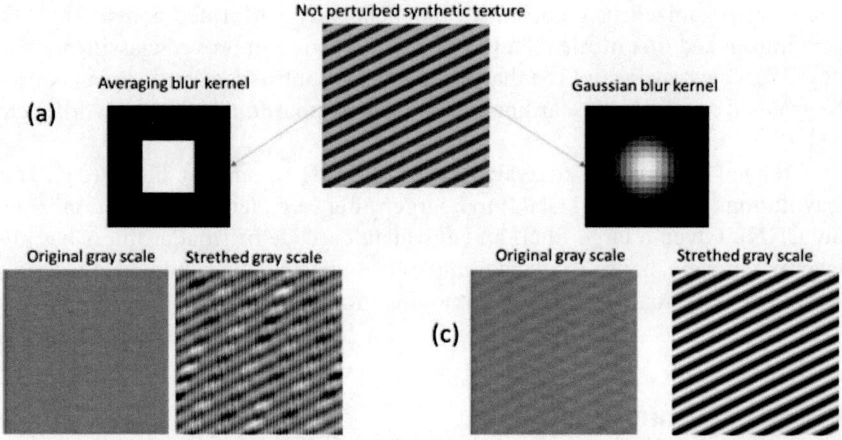

Fig. 2 a) Unperturbed synthetic texture, averaging (left) and Gaussian (right) perturbing kernels; upon perturbation using the kernels in (a), synthetic textures with (b) and without (c) aliasing effect

from the center are weighted by a bell-curve. Gaussian filters are less likely to lead to the same degree of aliasing effects as averaging filters (Fig. 2).

Second, we improve the synthetic texture library by diversifying it. As texture is a local image property associated to repeating local patterns, automated texture analysis needs to deal with texture patterns of different sizes. Hence, the synthetic textures used in this paper span four different sizes, 38×38, 46×46, 54×54, 62×62. This more closely mimics real-life textures, and the CNNs trained on them will have enhanced generalization capabilities. Overall, the training data consist of 1,490,760 texture images, the validation data of 745,380 texture images, and the test data of 745,380 texture images. Another improvement results from the broader range of bar period and thickness combinations used.

Layer	SN1	SN2	SN3	SN4	DN1	DN2	DN9	DN10	GN1	GN2	GN3	GN4
1	180x17x17 (all)	90x17x17 (all)	180x13x13 (all)	90x13x13 (all)	16x17x17 (all)	16x17x17 (all)	16x17x17 (all)	16x17x17 (all)	180x3x3 (linear)	180x3x3 (linear)	180x5x5 (linear)	180x5x5 (linear)
2	global max	global max	global max	global max	2x2 max	2x2 max	2x2 max	2x2 max	2x2 average	3x3 average	2x2 average	180x5x5 (linear)
3	0.25	0.25	0.25	0.25	0.25	0.25	0.25	0.25	180x3x3 (linear)	180x3x3 (linear)	180x5x5 (linear)	180x5x5 (linear)
4	180 (softmax)	180 (softmax)	180 (softmax)	180 (softmax)	16x5x5 (all)	32x5x5 (all)	32x5x5 (all)	32x5x5 (all)	2x2 average	3x3 average	2x2 average	180x5x5 (all)
5					2x2 max	2x2 max	2x2 max	2x2 max	180x3x3 (linear)	180x3x3 (all)	180x5x5 (all)	global max
6					0.25	0.25	0.25	0.25	2x2 average	global max	global max	0.25
7					90x3x3 (all)	90x3x3 (all)	64x3x3 (all)	64x3x3 (all)	180x3x3 (all)	0.25	0.25	180 (softmax)
8					global max	global max	global max	global max	global max	180 (softmax)	180 (softmax)	
3					0.25	0.25	0.25	0.25	0.25			
10					90 (all)	90 (all)	90 (all)	128 (all)	180 (softmax)			
11					0.5	0.5	0.5	0.5				
12					180 (softmax)	180 (softmax)	180 (softmax)	180 (softmax)				

- convolution - grouped convolution - pooling - dropout - fully connected - output

Fig. 3 Detailed schematic of all the CNN architectures used in this paper. For a given convolutional or fully connected layer, the specific activation function used is reported in parenthesis ("all" implies all activation functions were tested).

3.2 Novel CNN Architectures and Additional Activation Functions

We study twelve CNNs, some previously designed [2], whose architecture is shown in Fig. 3.

The CNNs were implemented using the TensorFlow library (Keras module) with a Python interface (Python version 3.10.5, TensorFlow version 2.7.1) [18]. All CNN architectures contain a global max pooling layer to enable a wide range of input image sizes, as opposed to only one image size. CNNs consist of one or several convolutional layers, generally followed by a pooling layer. Each convolutional layer is effectively a set of feature extraction filters. Note that, as in our previous work, we consistently employ convolutions with no padding [2]. The reason is that padding can cause artifacts at the image edge, likely leading to the detection of wrong texture directions. In the absence of padding, a sequence of convolutions will always result in a feature map of smaller size. Quantitatively, given a convolutional layer consisting of filters of size $(k \times l)$ without padding, an input image of size $(n \times m)$ will produce an output image of size $((n - k + 1) \times (m - l + 1))$. In addition, a pooling operation of dimensions $(o \times p)$ on an image of dimensions $(n \times m)$ will produce an image of dimensions $((n/o) \times (m/p))$. A sequence of K convolutions operations can be formally defined as follows.

$$I_{out}[\overline{n}, \overline{m}] = F_K * F_{K-1} * \ldots * F_2 * F_1 * I_{out}[n, m]. \tag{1}$$

In Eq. 1, K is the number of convolutions (with or without pooling layers), and I_{in} and I_{out} are the input and output images with respective sizes $(n \times m)$ and $(\overline{n} \times \overline{m})$. As pointed out, $n > \overline{n}$ and $m > \overline{m}$. Based on Eq. 1, two sequences of convolutions are size-equivalent if, given same size input images, the output images are also of the same size. The CNNs described in this paper always utilize a global max pooling layer after the last convolutional one. This enables processing input images of variable size. Since the minimum allowed image size for a global max pooling layer is 1×1, the minimum input image size for our CNNs is constrained accordingly. Hence, size-equivalent CNNs are also defined as having the same minimum allowed input image size. The expectation is that size-equivalent CNNs can be trained to have similar prediction performances, hence achieving the equivalence between cascades of filters and single large filters discussed in the Introduction. Here, network GN1 is size-equivalent to the networks DN1 and DN2 from our previous work, and network GN4 is size-equivalent to SN1 [2].

The role of the activation functions is critical for the CNN performances. Depending on properties such as boundaries, symmetry, and linearity, given activation functions will work better with specific CNNs architectures. We tested five additional activation functions available with Tensorflow 2.7.1, not included in our previous work: linear (ln), exponential (ex), Swish (sw), Gaussian error linear unit (ge), and hard sigmoid (hs). The previously employed activation functions, still used here, are: exponential linear unit (el), rectified linear unit (re), scaled exponential linear unit (se), sigmoid (si), soft plus (sp), soft sign (ss), and hyperbolical tangent (th).

4 Results and Discussion

The CNNs discussed in the Materials and Methods section were trained, validated and tested with the enhanced synthetic texture library. As in our previous work [2], we use the following directionality prediction error, which considers the inherent 180° periodicity.

$$error = \arccos(cos(|\alpha - \beta|)). \quad (2)$$

In Eq. (2) α is the true direction, and β is the predicted one.

Each CNN architecture was trained and validated in three replicates using Keras libraries and functions [19]. The training and validation curves show consistency across replicates. Good convergence behavior is evident, especially for the best-performing architectures, based on the categorical cross entropy (loss function) and RMSE. Subsequently, all CNNs architecture were tested. The metrics to evaluate CNN test performances are root mean square error (RMSE) and maximum error (MxE), both based on Eq. 2. RMSE characterizes the average performance of a CNN, MxE reflects the presence of large errors, not acceptable in many applications. The tests yield novel insight into CNN performance, discussed as follows. First, our data clearly show major improvements in CNN performances when trained on the

Table 2 MxE and RMSE data for the best shallow network SN1 trained, validated and tested on both the old and the enhanced synthetic texture library.

Training, validation and test on the old synthetic texture library												
Activation	el	se	si	sp	re	ss	th					
MxE	89	90	90	87	90	84	82					
RMSE	2.1	2.2	3.1	3.2	3.6	4.3	6.3					
Training, validation and test on the enhanced synthetic texture library												
Activation	ss	si	th	sp	se	ln	sw	el	ge	hs	ex	re
MxE	3	3	4	4	5	5	5	6	6	7	10	52
RMSE	0.28	0.29	0.32	0.34	0.32	0.34	0.40	0.32	0.37	0.33	0.74	0.74

Table 3 RMSE and MxE data for the best deep network DN2 trained, validated and tested on the enhanced synthetic texture library.

Activation	el	ss	sp	se	th	re	ln	ge	sw	si	ex	hs
MxE	90	90	90	90	90	90	90	90	90	90	90	90
RMSE	2.7	4.0	4.1	4.5	5.7	6.3	11.6	18.5	20.9	51.9	52.0	52.0

enhanced synthetic texture library. An instance is the performance of the best shallow CNN SN1 (see Table 2).

This is probably due to the absence of aliasing effects in the enhanced synthetic texture library, which improves the quality of the training and increases the generalization capability of the CNN. The presence of multi-size training data might also force the CNN to deal with more diverse texture patterns, which yields enhanced generalization capability.

The data also show that shallow CNNs have now a clearly better performance than deep CNNs, unlike our past results [2] (Table 3).

This is because the meaningful information about texture directionality probably lies in the lower-level features, which are collected in the first convolutional layer. Subsequent layers in the deep CNNs describe higher-level texture abstractions, such as relative location, which are less meaningful. This is more evident based on the data from the enhanced texture library, whereas it was not so clear from our previous data, due to the presence of aliasing effects. Additional tests (data not shown) were performed on deep CNNs with different number of filters (i.e., doubling in each subsequent layer), with no evidence of performance improvement. Tests on shallow CNNs with different sizes were also performed, but SN1 remains the best performing architecture. Additional tests focused on grouped CNNs, trained on the enhanced synthetic texture library. They span size-equivalent groups from the best shallow CNN, SN1, to the best deep CNN, DN2, from our previous work [2]. The best-performing grouped CNN is GN4 (Table 4).

Note that, as far as size-equivalence, GN4 is closest to the best shallow network SN1. Both CNNs are probably close to the optimal size for texture directionality detection when employing our training data.

Table 4 RMSE and MxE data for the grouped networks GN1, GN2 and GN4 trained, validated and tested on both the old and the enhanced synthetic texture library.

GN1	Activation	ge	sw	el	se	sp	si	th	ss	hs	ex	re	ln
	MxE	90	90	90	90	90	90	90	90	90	90	90	90
	RMSE	7.35	7.52	7.77	8.36	8.53	8.89	9.08	9.33	10.1	11.6	21.1	23.5
GN2	Activation	ge	sw	el	se	si	hs	th	sp	ss	ex	re	ln
	MxE	90	90	90	90	90	90	90	90	90	90	90	90
	RMSE	5.92	6.10	6.26	6.65	7.17	7.43	7.70	7.84	8.02	10.1	20.3	22.1
GN4	Activation	se	el	ge	sw	si	ln	ex	sp	th	ss	hs	re
	MxE	4	4	4	5	5	7	10	11	29	59	77	89
	RMSE	0.31	0.33	0.34	0.30	0.56	0.24	0.82	0.87	0.42	0.51	0.91	2.57

Notably, the best performing activation functions for the shallow network SN1 are different from our previous work (Table 2). Smooth rectifier activation functions such as ELU (el) and SELU (se) were previously reported to perform best [2]. With the enhanced synthetic texture library, mainly symmetric and bonded activation functions perform better, such as soft sign (ss), sigmoid (si), and hyperbolical tangent (th). This underscores the importance of the activation functions in the CNN performance.

Another important result is related to GN4, the best grouped CNN. In GN4, the linear activation function (ln), which is not bounded, is used in all convolutional layers except for the last one. In the last convolutional layer, smooth rectifier activation functions are used, such as SELU (se), as they yield the best performance. Overall, GN4 performance improves with non-bounded activation functions. This behavior is opposite to the best shallow CNN, SN1, where bounded activation functions such as SOFTSIGN (ss) perform best, probably due to their capability to counter the effect of outliers. While surprising at first, one explanation is that in GN4 non-bounded activation functions contribute with their properties to creating the conditions under which the cascade of smaller filters becomes equivalent to the unique larger filter of SN1 (cfr., Related Work section). Note that GN4, unlike the other grouped CNNs tested here, does not have any pooling operation between the convolutional layers. The absence of pooling operations could also contribute to the equivalence between the cascade of smaller filters of GN4 and the unique large filter of SN1.

5 Conclusions and Future Work

This effort has contributed to further elucidate the feasibility of CNNs trained on synthetic textures in the context of texture directionality detection. The employed synthetic textures closely resemble real-life ones, and hence the designed CNNs seem to nicely generalize to real-world images [2]. Overall, the best performing CNN is still the shallow SN1.

One important conclusion is that the quality of the training data is of fundamental importance. As evident from our previous results, the synthetic textures defined in the digital domain are associated with unwanted perturbation effects (i.e., aliasing effects), which may lead to the improper labeling of the training data. We mitigated adverse perturbation effects by employing enhanced perturbation procedures (i.e., Gaussian filters) and by ensuring no leftover aliasing effect is included in the training data. As a consequence, the performance of shallow CNNs clearly improved.

Another conclusion is that the better performance of shallow CNNs, when compared to deep CNNs, suggests that shallow CNNs probably enable the full characterization of texture directionality in the unique convolutional layer employed. Hence, there is probably no need to resort to deeper architectures for increased accuracy.

Lastly, our tests suggest that grouped CNNs, rather than deep CNNs, might be viable alternatives to shallow CNNs. This is possibly associated to the conceptual equivalence between cascades of smaller filters and large ones. Such equivalence seems to be suggested by the very similar performances of size-equivalent grouped and shallow CNNs (cfr., Materials and Methods).

The implementation of the described methods is available in an open Github repository [20].

Based on the above, one future work direction is to test additional grouped CNNs, consisting of a smaller or larger number of layers. As our intuition is that size-equivalent shallow and grouped CNNs might perform similarly in texture directionality detection, the new architectures should span as many size-equivalent shallow networks as feasible, including the best one, SN1. Another future work direction is related to the absence of pooling operations in the best-performing grouped network GN4. Whereas this might be evidence that such an absence contributes to the equivalence between the cascade of smaller filters in GN4 and the unique larger filter in shallow networks, it is still worth testing different CNN architectures with no pooling operations, including deep ones. This would provide a more complete picture of the performance of CNNs in texture directionality detection, possibly leading to further enhancements of the proposed approach.

Disclaimer

Commercial products are identified in this document in order to specify the experimental procedure adequately. Such identification is not intended to imply recommendation or endorsement by the National Institute of Standards and Technology, nor is it intended to imply that the products identified are necessarily the best available for the purpose.

References

1. Julesz, B., Gilbert, E.N., Shepp, L.A., Frisch, H.L.: Inability of humans to discriminate between visual textures that agree in second order statistics: revisited. Perception **2**(4), 391–405 (1973). https://doi.org/10.1068/p020391
2. Kociołek, M., Kozłowski, M., Cardone, A.: A convolutional neural networks-based approach for texture directionality detection. Sensors **22**, 5620 (2022). https://doi.org/10.3390/S22020562
3. Hassekar, P., Sawant, R.R.: Experimental analysis of perceptual based texture features for image retrieval. In: Proceedings - 2015 International Conference on Communication, Information and Computing Technology, ICCICT 2015, pp. 1–6 (2015). https://doi.org/10.1109/ICCICT.2015.7045666
4. Lin, X., Ye, L., Zhong, W., Zhang, Q.: Directionality-based modified coefficient scanning for image coding. In: Proceedings - 2015 8th International Congress on Image and Signal Processing, CISP 2015, vol. 1, no. CISP, pp. 194–198 (2016). https://doi.org/10.1109/CISP.2015.7407874
5. Maskey, M., Newman, T.S.: On measuring and employing texture directionality for image classification. Pattern Anal. Appl. (2021). https://doi.org/10.1007/s10044-021-01013-8
6. Lin, M., et al.: Longitudinal assessment of carotid plaque texture in three-dimensional ultrasound images based on semi-supervised graph-based dimensionality reduction and feature selection. Comput. Biol. Med. **116**, 103586 (2020). https://doi.org/10.1016/J.COMPBIOMED.2019.103586
7. Edachery, V., Shashank, R., Kailas, S.V.: Influence of surface texture directionality and roughness on wettability, sliding angle, contact angle hysteresis, and lubricant entrapment capability. Tribol. Int. **158**, 106932 (2021). https://doi.org/10.1016/J.TRIBOINT.2021.106932
8. Kempton, D.J., Ahmadzadeh, A., Schuh, M.A., Angryk, R.A.: Improving the functionality of tamura directionality on solar images. In: Proceedings - 2017 IEEE International Conference on Big Data, Big Data 2017, vol. 2018-Janua, pp. 2518–2526 (2017). https://doi.org/10.1109/BigData.2017.8258210
9. Feng, D., Li, C., Xiao, C., Sun, W.: Research of spectrum measurement of texture image. In: World Automation Congress (WAC), pp. 163–165 (2012)
10. Padlia, M., Sharma, J.: Fractional sobel filter based brain tumor detection and segmentation using statistical features and SVM. In: Nath, V., Mandal, J.K. (eds.) Nanoelectronics, Circuits and Communication Systems. LNEE, vol. 511, pp. 161–175. Springer, Singapore (2019). https://doi.org/10.1007/978-981-13-0776-8_15
11. Rasband, W.S., et al.: ImageJ. U.S. National Institutes of Health, Bethesda, Maryland, USA (1997). https://imagej.nih.gov/ij/. Accessed 08 Nov 2021
12. Jafari-Khouzani, K., Soltanian-Zadeh, H.: Radon transform orientation estimation for rotation invariant texture analysis. IEEE Trans. Pattern Anal. Mach. Intell. **27**(6), 1004–1008 (2005). https://doi.org/10.1109/TPAMI.2005.126
13. Kociolek, M., Bajcsy, P., Brady, M., Cardone, A.: Interpolation-based gray-level co-occurrence matrix computation for texture directionality estimation. In: Signal Processing - Algorithms, Architectures, Arrangements, and Applications Conference Proceedings, SPA, vol. 2018-September, pp. 146–151 (2018). https://doi.org/10.23919/SPA.2018.8563413
14. Krizhevsky, A., Sutskever, I., Hinton, G.E.: Imagenet classification with deep convolutional neural networks. Commun. ACM **60**(6), 84–90 (2017)
15. Xie, S., Girshick, R., Dollár, P., Tu, Z., He, K.: Aggregated residual transformations for deep neural networks. In: Proceedings - 30th IEEE Conference on Computer Vision and Pattern Recognition, CVPR 2017, vol. 2017-January, pp. 5987–5995 (2017) https://doi.org/10.1109/CVPR.2017.634
16. Fragoso, V., Srivastava, G., Nagar, A., Li, Z., Park, K., Turk, M.: Cascade of box (CABOX) filters for optimal scale space approximation. In: IEEE Computer Society Conference on Computer Vision and Pattern Recognition Workshops, pp. 126–131 (2014). https://doi.org/10.1109/CVPRW.2014.24

17. Gonzalez, R., Woods, R.: Digital Imaging Processing, 4th edn. Pearson, Boston (2022). Accessed 01 Feb 2023. https://www.pearson.com/en-us/subject-catalog/p/digital-image-processing/P200000003224/9780137848560
18. Abadi, M., et al.: TensorFlow: a system for large-scale machine learning. In: 12th USENIX Symposium on Operating Systems Design and Implementation (OSDI 2016), pp. 265–283 (2016). https://www.usenix.org/conference/osdi16/technical-sessions/presentation/abadi
19. Gulli, A., Pal, S.: Deep Learning with Keras. Packt Publishing Ltd., Birmingham (2017)
20. Kociolek, M., Cardone, A.: Marcinkociolek/DirDetectCORES (2023). https://github.com/marcinkociolek/DirDetectCORES2023

Knowledge Distillation with Relative Representations for Image Representation Learning

Patrick Ramos, Raphael Alampay, and Patricia Abu

Abstract Relative representations allow the alignment of latent spaces which embed data in extrinsically different manners but with similar relative distances between data points. This ability to compare different latent spaces for the same input lends itself to knowledge distillation techniques. We explore the applicability of relative representations to knowledge distillation by training a student model such that the relative representations of its outputs match the relative representations of the outputs of a teacher model. We test our Relative Representation Knowledge Distillation (RRKD) scheme on supervised and self-supervised image representation learning with MNIST and show that an encoder can be compressed to 47.71% of its original size while maintaining 91.92% of its full performance. We demonstrate that RRKD is competitive with or outperforms other relation-based distillation schemes in traditional distillation setups (CIFAR-10, CIFAR-100, SVHN) and in a transfer learning setting (Stanford Cars, Oxford-IIIT Pets, Oxford Flowers-102). Our results indicate that relative representations are an effective signal for knowledge distillation. Code is made available at https://github.com/Ramos-Ramos/rrkd.

Keywords Knowledge distillation · Relative representations · Latent space

1 Introduction

Although the latent spaces learned by neural networks are expected to be solely reliant on their data and optimization constraints, stochastic factors such as random weight initialization cause models with similar constraints to learn different latent spaces. However, these spaces are actually intrinsically similar and differ by a quasi-isometric transformation [15]. Relative representations [12] leverage this phenomenon and re-express latents of the same input but in different spaces as vectors of similarities to other "anchor" latents in their respective vector spaces. This method of representing

P. Ramos (✉) · R. Alampay · P. Abu
ALIVE, Ateneo de Manila University, Quezon City, Philippines
e-mail: patrick.john.ramos@obf.ateneo.edu

R. Burduk et al. (eds.), *Progress on Pattern Classification, Image Processing and Communications*, Lecture Notes in Networks and Systems 766,
https://doi.org/10.1007/978-3-031-41630-9_14

latents is invariant to isometry and allows the alignment of latent spaces for zero-shot communication between them.

The idea of having separate models trained on the same task with the same data draws similarities with knowledge distillation [6], where a smaller student model must generalize to data in a manner similar to a larger teacher one. This is usually done by using one or more outputs of the teacher model (e.g. final class logits, intermediate features) as a supervisory signal for training the student.

There are two factors driving the intuition that relative representations can be used for knowledge distillation. Firstly, [12] report that in a collection of models optimized with the same objective across different hyperparameters, model performance is correlated with latent space similarity to a reference gold model. Secondly, computing relative representations is fully differentiable, allowing their similarity to be used as a knowledge distillation objective.

We conduct a preliminary but extensive investigation exploring the applicability of relative representations to knowledge distillation in an image representation learning setting. Our Relative Representation Knowledge Distillation (RRKD) method consists of converting student and teacher outputs to relative representations and optimizing a matching objective between them as opposed to the original representations. Our contributions are as follows:

- We design a knowledge distillation scheme, called RRKD, centered around matching the relative representations of a student to those of a teacher.
- In both supervised and self-supervised image representation learning experiments with MNIST, we demonstrate that relative representations are capable of distilling knowledge in a teacher-student framework.
- We show that RRKD can outperform similar relation-based distillation methods across a variety of benchmarks, with results extending to transfer learning.
- Through an ablation study, we show that online selection of anchors by using in-batch references is an effective anchor selection strategy.

2 Related Work

Knowledge distillation [6] is a model compression method that trains a small student network to mimic the behavior of a large reference teacher model. Rather than rely solely on supervision from the training data, students learn from targets created by the teacher. The classical form of knowledge distillation is seen in the classification setting, where the student's logits are matched to "soft target" logits generated by the teacher alongside the ground truth hard labels from the training data.

FitNets [20] improve on classical distillation by introducing feature representations from intermediate layers as targets. When using intermediate representations during distillation, the outputs of "guided" layers in the student are matched to those of "hint" layers in the teacher. Intermediate feature-based distillation methods may also distill transformations of the intermediate features such as attention maps [24] or neuron selectivities [7] computed from CNN feature maps.

While knowledge distillation methods typically distill the final or intermediate encodings of inputs, there exist several relation-based distillation methods that instead distill the relationships between data points, like our proposed method. Such methods distill the relationships between triplets or pairs of data. Prior works have formulated these relationships as distances in embedding space [16, 23], instance relation graphs [11], joint probability distributions [18], and correlations [19]. Most similar to our work is similarity-preserving knowledge distillation [22], which distills the l_2-normalized outer product of a batch of model activations with itself. We also use the outer product of model activations as a distillation signal, but perform l_2-normalization before the outer product to create a cosine similarity map of instances within a batch, as discussed in Sect. 4.

3 Background: Relative Representations

Given training data X and an encoder $\phi : X \rightarrow \mathbb{R}^d$, the standard d-dimensional embedding $\phi(x) \in \mathbb{R}^d$ for $x \in X$ is referred to as the absolute representation of x. Relative representations instead express x in terms of its similarity to, or *relative* to, other points in X. Specifically, x is represented as an m-dimensional vector of similarities of its absolute representation to the absolute representations of a set of m anchor points.

To create a relative representation, we select a set of m anchor points from the training data, denoted as $A \subseteq X$. The relative representation of $x \in X$ is then computed as

$$r(x) = \langle s(\phi(x), \phi(A_0)), \ldots, s(\phi(x), \phi(A_{m-1})) \rangle \in \mathbb{R}^m, \tag{1}$$

where $s(\cdot, \cdot)$ is a similarity function producing a scalar similarity score for two latents. This effectively projects x to an m-dimensional space where each dimension is its similarity to one of the m anchors.

4 Relative Representation Distillation

Figure 1 provides an overview of RRKD. Given a batch of n inputs, we extract teacher and student absolute representations $Z_t, Z_s \in \mathbb{R}^{n \times d}$ using a frozen teacher encoder ϕ_t and a learnable student encoder ϕ_s, respectively. The teacher and student absolute representations are then converted to relative representations $V_t, V_s \in \mathbb{R}^{n \times n}$. Computing relative representations for zero-shot communication between latent spaces typically involves choosing a set of anchors beforehand to use in computing all relative representations. However, instead of selecting a fixed set of anchor points, we use all embeddings in the batch as anchors akin to the intra-batch comparisons performed

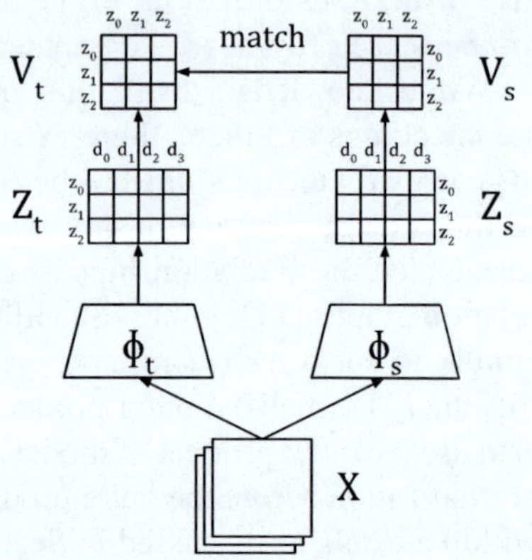

Fig. 1 Relative representation knowledge distillation overview. Teacher ϕ_t and student ϕ_s respectively encode inputs X into latents Z_t and Z_s. The latents are converted to relative representations V_t and V_s using themselves as anchors. A cosine similarity-based loss is used to encourage V_s to match V_t. ϕ_t is frozen.

in contrastive learning [3]. This anchor selection method is compared to other strategies in an ablation study in Sect. 5.4. We follow [12] and choose cosine similarity as the similarity function s, resulting in V_t and V_s being $n \times n$ cosine similarity maps of embeddings to each other. The relative representations can be formulated as

$$V_m = \widetilde{Z_m}\widetilde{Z_m}^\intercal, \tag{2}$$

where $\widetilde{Z_m}$ is a row-wise l_2-normalized Z_m and $m \in \{t, s\}$ refers to the teacher (t) and student (s) models. This computes the cosine similarity map by taking the outer product of the l_2-normalized absolute representations.

The distillation loss $\mathcal{L}_D$ then encourages the student's relative representations to match those of the teacher's by taking the negative average logarithm of their pairwise cosine similarities, which are rescaled between 0 and 1. We add a small constant $\epsilon =$ 1e$-$8 to the rescaled cosine similarity before the logarithm for numerical stability. The use of cosine similarity is based on experiments by [12] using the similarity of a network's relative representations with those of a reference model as a performance proxy.

$$\mathcal{L}_D = -\frac{1}{n}\sum_{i=0}^{n-1} \log\left(\frac{V_{t,i} \cdot V_{s,i} + 1}{2\|V_{t,i}\|_2\|V_{s,i}\|_2} + \epsilon\right) \tag{3}$$

$\mathcal{L}_D$ assumes that the networks ϕ_t and ϕ_s are encoders that project to a latent space, such as models trained with self-supervised learning, rather than supervised classifiers that produce class logits. While the logit space can be considered a latent space where the dimensions correspond to each class, permutation symmetries in such a space may result in student relative representations that are similar to a teacher's but assign the highest scores to the wrong class. Therefore it is not recommended to use the logits as the latent space for computing relative representations.

Nevertheless, it is still possible to distill a teacher classifier into a student encoder using relative representations. Simply removing the teacher's classification head results in an encoder. The distillation then happens in the latent space prior to the projection to class logits. Furthermore, class labels can be used in distillation if available. When distilling a model with labels Y, we add a classification head on top of ϕ_s which projects Z_s into class label predictions P and add the classical cross entropy classification loss $\mathcal{L}_{CE}$ to the total loss. The loss for distilling models with labels is then a linear combination of the distillation and cross entropy losses:

$$\mathcal{L}_L = \lambda_D \mathcal{L}_D + \lambda_{CE} \mathcal{L}_{CE}(P, Y), \tag{4}$$

where λ_D and λ_{CE} are weights for the distillation and cross entropy losses respectively. In practice, we set $\lambda_D = 1$ and use $\lambda_{CE} = 1$ when distilling with class labels and $\lambda_{CE} = 0$ when distilling without. Setting $\lambda_{CE} = 0$ makes the loss function equivalent to Eq. 3.

Note that the use of relative representations always results in latents of the same size between the teacher and student, which frees us from the usual restriction in knowledge distillation of needing the same dimensionality for teacher and student representations.[1] This means not only can student models be "shorter" (less layers) than the teacher, but they can also be "thinner" (smaller dimension size).

5 Experiments

We evaluate RRKD in an image representation learning setting. We perform distillation experiments for image encoders trained with supervision and self-supervision, compare RRKD to other relation-based distillation schemes, and perform an ablation study on the method of selecting anchors. Note that in all experiments, the aim is not to maximize raw performance on the target dataset but to maximize the preservation of a teacher's performance.

5.1 Distillation Compared to Training from Scratch

We distill both self-supervised and supervised MLP encoders trained on MNIST using RRKD. We evaluate all models with linear probe evaluation using a logistic regression classifier. This includes models distilled with class labels, as our primary goal is to use RRKD to train student encoders rather than classifiers, and the standard

[1] Some works try to address this by projecting the student outputs with a learnable linear layer to have the same dimensionality as the teacher [8, 20]. Our work is similar in that regard as computing the relative representations can also be considered a linear projection, but our method does not require learning the projection weights.

protocol for evaluating encoder representations is linear probe evaluation [1, 3]. Furthermore, we find it more effective to discard the classification head after distillation and attach a new linear head.

To investigate the value of distillation compared to simply training the smaller model from scratch, students are compared to baselines which follow the same architecture but are trained in the same manner as the teacher. Following a hyperparameter sweep across learning rates {1e−1, 1e−2, 1e−3} (extended until 1e−8 for the self-supervised baseline), we train all models for 20 epochs using SGD with learning rate 1e−1 and batch size 128, except for the self-supervised baseline, which was trained with learning rate 1e−8. The models in the self-supervised MNIST encoder distillation experiment use 0.9 momentum to accelerate optimization.

Self-supervised MNIST. We train an MLP auto-encoder with hidden layer sizes {128, 64, 128} on MNIST with an MSE pixel reconstruction loss. We use the encoder of the auto-encoder as the teacher and distill it into a student MLP with layer sizes {64, 32}. This student can be viewed as the encoder of an auto-encoder with hidden layer sizes {64, 32, 64}, which serves as the baseline. All models use ReLU activation and 0.5 dropout. Distillation is performed using the standard relative representation loss described in Eq. 3.

We report linear probe evaluation accuracies averaged across three trials and model parameter counts in Table 1. With only 47.71% of the number of parameters of the teacher, the student is able to retain 91.92% of the teacher's performance. The student also achieves a higher accuracy than a baseline trained from scratch, outperforming it by 6.74%.

Supervised MNIST. We train an 3-layer MLP with hidden layer sizes {1200, 1200} on MNIST classification. Stripping the classification layer leaves a 2-layer MLP teacher encoder projecting to a 1200-dimensional latent space. The student is a 2-layer MLP that has one hidden layer of size 32 and projects to 32 dimensions. ReLU activation and dropout of 0.5 are also used. Relative representation distillation is performed using class labels with the loss described in Eq. 4.

The average accuracies over three trials and model sizes are presented in Table 2. The student preserves 97.18% of the teacher's accuracy while being merely 1.3% of its size. Furthermore, the student outperforms the baseline by 2.07%.

These MNIST experiments demonstrate that transferring teacher knowledge with RRKD is capable of producing compressed encoders that are more performant than ones that were trained from scratch.

Table 1 Distillation results for self-supervised MNIST models.

Model	Parameter Count	Linear
Teacher (AE-64)	109K	87.99
Baseline (AE-32)	52K	74.14
Student (AE-32)	52K	80.88

Table 2 Distillation results for supervised MNIST models.

Model	Parameter Count	Linear
Teacher (MLP-1200)	2M	95.64
Baseline (MLP-32)	26K	90.87
Student (MLP-32)	26K	92.94

5.2 Comparison to Other Relation-Based Distillation Methods

We compare our relative representation distillation scheme to SPKD [22] and LPKD [2], two other relation-based distillation methods. SPKD distills a similarity map computed by the row-wise l_2-normalized outer product of latents from a teacher hint layer to that of a student guided layer, and can be combined with a cross entropy loss with class labels. In our experiments, we use $\gamma = 1$ and follow the SPKD authors by only using the last hidden layers as hint and guided layers. Meanwhile, LPKD distills a squared error distance map and was proposed to be combined with a cross entropy loss with class labels and a soft cross entropy loss with teacher logits. We use $\gamma = 1.5$, $\lambda = 2$, and $\tau = 0.5$. For both methods, please refer to [22] and [2] for an explanation of these hyperparameters.

Across CIFAR-10 [10], CIFAR-100 [10], and SVHN [13], we train a supervised ResNet-18 [5] teacher with a final hidden layer of 512. The teacher is then distilled using the distillation losses and class labels into a ResNet-9 encoder with a 256 embedding size, which we refer to as ResNet-9 $\times$ 0.5. Images are kept at their original 32×32 resolution and are augmented only with random horizontal flipping. We train and distill for 50 epochs on CIFAR-10 and CIFAR-100 and for 20 epochs on SVHN. Models are optimized with SGD with learning rate 1e$-$1 (determined after a hyperparameter sweep across learning rates {1e$-$1, 1e$-$2, 1e$-$3}) and batch size 128. Momentum of 0.9 is used during distillation to be consistent with the use of momentum across benchmark distillation methods. These hyperparemters are also used in subsequent experiments.

Table 3 Comparison to other relation-based distillation methods. Best non-teacher results are indicated in **bold**.

Method/Model	Parameter Count	CIFAR-10 [10]	CIFAR-100 [10]	SVHN [13]
Teacher	11M	88.13	60.72	94.66
SPKD [22]	1M	85.85	56.35	**94.56**
LPKD [2]	1M	85.56	57.04	94.49
RRKD (ours)	1M	**86.21**	**57.48**	94.55

Table 4 Comparison to other relation-based distillation methods on transfer learning with distillation.

Method/Model	Parameter Count	Stanford Cars [9]	Oxford-IIIT Pets [17]	Oxford Flowers-102 [14]
Teacher	21M	78.60	90.52	85.76
SPKD [22]	11M	72.63	77.19	80.50
LPKD [2]	11M	70.47	73.81	77.27
RRKD (ours)	11M	**76.79**	**81.22**	**84.62**

Table 5 Anchor selection ablation study results.

Anchor Selection Method	Anchors	MNIST Test Accuracy
In-batch	128	**81.42**
Random	128	80.21
Per class	130	80.59
Best per class	130	80.52

Test accuracies on each dataset is reported in Table 3. RRKD outperforms baseline methods on CIFAR-10 (+0.36% from next best method) and CIFAR-100 (+0.44%), and has only a small deficit (−0.01%) compared to the best performing method on SVHN, all while being 11.05% the size of the teacher on average. These results indicate that RRKD can perform on par with other relation-based distillation methods if not better.

5.3 *Transfer Learning*

We also conduct distillation experiments in the transfer learning setting, where a pretrained student is fine-tuned while being distilled from a pretrained teacher already fine-tuned on the target dataset. Teachers are ResNet-34 s pretrained on ImageNet [4] and fine-tuned on Stanford Cars [9], Oxford-IIIT Pets [17], and Oxford Flowers-102 [14]. Students are ImageNet-pretrained ResNet-18 s. Images are augmented with Inception-style random cropping [21] to 224×224, horizontal flipping, and color jitter. We train and distill models for 20 epochs on Stanford Cars and Oxford-IIIT Pets and for 50 epochs on Oxford Flowers-102.

We report the accuracies of the distilled models in Table 4. RRKD outperforms other methods across all datasets. Averaged across the three datasets, the distilled student is capable of retaining 95.36% of the teacher's original accuracy with 52.64% the parameter count of a full teacher, showing that distilling relative representations can also be extended to transfer learning.

5.4 *Anchor Selection*

We perform an ablation study on the anchor selection method. We evaluate the linear probe evaluation of encoders distilled from the self-supervised MNIST MLPs described in Sect. 5.1 using four different anchor selection strategies:

- **In-batch.** This is the main method used in this study, described in Sect. 4. We use all embeddings in the batch as anchors, with a batch size of 128.
- **Random.** We select a fixed set 128 of random embeddings to be used as anchors throughout training.
- **Per class.** We randomly sample 13 examples per class to create a fixed set of 130 anchors. We choose 13 examples in order to have the smallest number of anchors greater than or equal to the number of in-batch anchors while giving each class equal representation.
- **Best per class.** This is similar to the previous method of choosing 13 examples per class but we instead choose the 13 instances per class that the teacher model is most confident in. For each class, we identify the 13 correctly classified examples with the largest probability for that particular class according to the teacher.

We show the linear probe evaluation results averaged across three trials in Table 5. Using in-batch anchors provides the highest linear probe evaluation, even when other anchor selection methods (per class, best per class) have more anchors. We hypothesize this may be because in-batch anchors expose the student to an larger number of reference points during training as using in-batch anchors means that the anchors change at each step and that all data points in the dataset are treated as anchors at some point.

6 Conclusion

We devise a knowledge distillation method based on relative representations, called RRKD. In image representation learning, student image encoders trained with RRKD are more performant than similar models trained without distillation. On a variety of image classification datasets, students trained with RRKD are competitive with if not better than other relation-based distillation methods.

Future works can further this study by exploring the scoring function used in generating relative representations alongside the loss function for comparing these representations. These experiments can also be extended to other architectures such as Transformer-based models and to other domains such as language and graph representation learning.

References

1. Caron, M., et al.: Emerging properties in self-supervised vision transformers. In: Proceedings of the IEEE/CVF International Conference on Computer Vision, pp. 9650–9660 (2021)
2. Chen, H., Wang, Y., Xu, C., Xu, C., Tao, D.: Learning student networks via feature embedding. IEEE Trans. Neural Netw. Learn. Syst. **32**(1), 25–35 (2020)
3. Chen, T., Kornblith, S., Norouzi, M., Hinton, G.: A simple framework for contrastive learning of visual representations. In: International Conference on Machine Learning, pp. 1597–1607. PMLR (2020)
4. Deng, J., Dong, W., Socher, R., Li, L.J., Li, K., Fei-Fei, L.: Imagenet: a large-scale hierarchical image database. In: 2009 IEEE Conference on Computer Vision and Pattern Recognition, pp. 248–255. IEEE (2009)
5. He, K., Zhang, X., Ren, S., Sun, J.: Deep residual learning for image recognition. In: Proceedings of the IEEE Conference on Computer Vision and Pattern Recognition, pp. 770–778 (2016)
6. Hinton, G., Vinyals, O., Dean, J., et al.: Distilling the knowledge in a neural network. arXiv preprint arXiv:1503.02531, vol. 2, no. 7 (2015)
7. Huang, Z., Wang, N.: Like what you like: Knowledge distill via neuron selectivity transfer. arXiv preprint arXiv:1707.01219 (2017)
8. Jiao, X., Yin, Y., Shang, L., Jiang, X., Chen, X., Li, L., Wang, F., Liu, Q.: Tinybert: distilling BERT for natural language understanding. arXiv preprint arXiv:1909.10351 (2019)
9. Krause, J., Stark, M., Deng, J., Fei-Fei, L.: 3D object representations for fine-grained categorization. In: Proceedings of the IEEE International Conference on Computer Vision Workshops, pp. 554–561 (2013)
10. Krizhevsky, A., Hinton, G., et al.: Learning multiple layers of features from tiny images (2009)
11. Liu, Y., Cao, J., Li, B., Yuan, C., Hu, W., Li, Y., Duan, Y.: Knowledge distillation via instance relationship graph. In: Proceedings of the IEEE/CVF Conference on Computer Vision and Pattern Recognition, pp. 7096–7104 (2019)
12. Moschella, L., Maiorca, V., Fumero, M., Norelli, A., Locatello, F., Rodolà, E.: Relative representations enable zero-shot latent space communication. arXiv preprint arXiv:2209.15430 (2022)
13. Netzer, Y., Wang, T., Coates, A., Bissacco, A., Wu, B., Ng, A.Y.: Reading digits in natural images with unsupervised feature learning (2011)
14. Nilsback, M.E., Zisserman, A.: Automated flower classification over a large number of classes. In: 2008 Sixth Indian Conference on Computer Vision, Graphics & Image Processing, pp. 722–729. IEEE (2008)
15. Olah, C.: Visualizing representations: deep learning and human beings (2015). http://colah.github.io/posts/2015-01-Visualizing-Representations/
16. Park, W., Kim, D., Lu, Y., Cho, M.: Relational knowledge distillation. In: Proceedings of the IEEE/CVF Conference on Computer Vision and Pattern Recognition, pp. 3967–3976 (2019)
17. Parkhi, O.M., Vedaldi, A., Zisserman, A., Jawahar, C.: Cats and dogs. In: 2012 IEEE Conference on Computer Vision and Pattern Recognition, pp. 3498–3505. IEEE (2012)
18. Passalis, N., Tefas, A.: Learning deep representations with probabilistic knowledge transfer. In: Proceedings of the European Conference on Computer Vision (ECCV), pp. 268–284 (2018)
19. Peng, B., Jin, X., Liu, J., Li, D., Wu, Y., Liu, Y., Zhou, S., Zhang, Z.: Correlation congruence for knowledge distillation. In: Proceedings of the IEEE/CVF International Conference on Computer Vision, pp. 5007–5016 (2019)
20. Romero, A., Ballas, N., Kahou, S.E., Chassang, A., Gatta, C., Bengio, Y.: Fitnets: hints for thin deep nets. arXiv preprint arXiv:1412.6550 (2014)
21. Szegedy, C., Liu, W., Jia, Y., Sermanet, P., Reed, S., Anguelov, D., Erhan, D., Vanhoucke, V., Rabinovich, A.: Going deeper with convolutions. In: Proceedings of the IEEE Conference on Computer Vision and Pattern Recognition, pp. 1–9 (2015)
22. Tung, F., Mori, G.: Similarity-preserving knowledge distillation. In: Proceedings of the IEEE/CVF International Conference on Computer Vision, pp. 1365–1374 (2019)

23. You, S., Xu, C., Xu, C., Tao, D.: Learning from multiple teacher networks. In: Proceedings of the 23rd ACM SIGKDD International Conference on Knowledge Discovery and Data Mining, pp. 1285–1294 (2017)
24. Zagoruyko, S., Komodakis, N.: Paying more attention to attention: Improving the performance of convolutional neural networks via attention transfer. arXiv preprint arXiv:1612.03928 (2016)

An Approach for CT Image Conversion Using Filtering Based on Quaternion Mathematics

Mahmoud Nasr, Adam Piórkowski, Krzysztof Brzostowski, and Fathi E. Abd El-Samie

Abstract Computed tomography (CT) images are widely used in medical examination applications. They are constructed from the raw data using different kernels. However, due to the technology used, the software finds problems with image segmentation in the presence of noise or blurred edges. In this paper, we use quaternion mathematics for image de-noising by applying the bilateral filtering algorithm to reconstruct a pure soft image from the images reconstructed with sharp kernels, without using the raw data. Our output is compared to reconstructed images with different denoising techniques, such as non-local means and wavelet denoising. The reconstructed images using other techniques are assessed using structural index similarity (SSIM) and peak signal-to-noise ratio (PSNR). Results show high similarity by comparing the outputs of sharp kernels to that of the B30f. Furthermore, the results show that the proposed work can efficiently increase the similarity between the images reconstructed with hard and soft kernels.

Keywords Quaternion · Bilateral Filter · Reconstruction Kernels · Denoising · SSIM · PSNR

M. Nasr (✉) · A. Piórkowski
Department of Biocybernetics and Biomedical Engineering, AGH University of Science and Technology, A. Mickiewicza 30 Av, 30-059 Cracow, Poland
e-mail: nasr@agh.edu.pl

M. Nasr
Engineering Mathematics and Physics Department, Faculty of Engineering and Technology, Future University in Egypt (FUE), New Cairo 11845, Egypt
e-mail: mahmoud.nasr@fue.edu.eg

K. Brzostowski
Faculty of Information and Communication Technology, Wroclaw University of Science and Technology, Wroclaw, Poland
e-mail: krzysztof.brzostowski@pwr.edu.pl

F. E. A. El-Samie
Department of Electronics and Electrical Communications Engineering, Faculty of Electronic Engineering, Manoufia University, Menouf, Egypt

R. Burduk et al. (eds.), *Progress on Pattern Classification, Image Processing and Communications*, Lecture Notes in Networks and Systems 766,
https://doi.org/10.1007/978-3-031-41630-9_15

1 Introduction

Imaging and diagnostic techniques in medical applications have seen tremendous and quick development and improvement over the recent decades, where these techniques have become pivotal in diagnosis of several diseases. Medical images are crucial for the information they give about heart, brain, nerves, etc., for acquiring an interior glimpse of the human body. In addition, a broad range of mathematical techniques can be used in medical imaging to examine whether the healthy parts of the body got infected or not. On the other hand, losing any particular data during image acquisition may end in a calamity, such as the patient's death. So the main obstacle in medical image acquisition is obtaining clear images without losing relevant information.

The image gets corrupted with noise or artifacts during the acquisition process and maybe in other processing stages. Noise is referred to as unwanted changes to the pixel values of the original image. This noise reduces the image visibility, affecting the small parts with low contrast. As a result, de-noising in imaging systems has become a fundamental stage in image pre-processing. The most common medical imaging techniques include CT [1, 2], Ultrasound (US) [1], and Magnetic Resonance (MR) imaging [1]. We will focus in our work on CT images.

Computed Tomography (CT) is a medical imaging technique that uses X-rays to create detailed images of the body. CT scans are especially helpful in identifying disorders affecting the head, chest, abdomen, and pelvis. The imaging system captures X-ray images from various angles, which are then combined by computer algorithms to create an image of the scanned body parts [3]. These X-ray projections are collected and saved as raw data, which means that the data needs enormous storage to keep them, but it is usually deleted after some days. Therefore, the reconstructed images are created using different kernels before deleting the raw data. These kernels are reconstructed using Siemens scanners that frequently employ two dedicated kernels: B46f (explicitly intended for reliable assessment of coronary calcifications and stents with 3D edge-preserving noise reduction techniques) and B26f (for tissue) [4, 5]. The B26f kernel provides a smooth definition of tissues with less noise, but it blurs the borders of high-density objects. Thus, the B46f kernel is employed to display them. However, noise in images rebuilt with finer kernels complicates segmentation. Unfortunately, the existence of noise in images reconstructed with sharper kernels makes segmentation challenging. As a result, in some instances, it is necessary to convert an image reconstructed with a hard kernel to one reconstructed with a soft kernel [6].

Bilateral filtering is a non-linear technique to smooth images, while keeping the edges. It is especially beneficial for image denoising, since it can eliminate noise while keeping crucial image details and edges. The bilateral filter compares each pixel to its surrounding pixels and adjusts its intensity depending on spatial distance and similarity. This enables the filter to selectively smooth portions of the image, while keeping edges and small features. In recent years, bilateral filtering has become a prominent image de-noising technique, and it is frequently utilized in applications such as digital photography, medical imaging, and computer vision [7]. Our work will focus on implementing the bilateral filter using quaternion mathematics to transform sharp kernels into soft kernels.

This paper is divided into five parts. Section 1 discusses the introduction. In Sect. 2, we will talk about the related work of previous techniques for medical image denoising. In Sect. 3, we introduce a brief introduction to bilateral filtering and quaternion mathematics. In Sect. 4, we present the proposed algorithm. Section 5 introduces the simulation results. Finally, Sect. 6 gives the concluding remarks and the suggested future work.

2 Related Work

Various post-processing noise-reduction technologies do not depend directly on raw data; they can partially address the above-mentioned problem, and be immediately applied to a reconstructed image [6]. Some examples are the linear smoothing filter, the bilateral filter, and the non-local means filter. Filtering was used, because standard CT images are contaminated with additive Gaussian noise after reconstruction [8]. This study is intended to investigate if employing denoising filters instead of rebuilding newer images from raw data allows to reconstruct an image with a hard kernel that seems as if it was reconstructed with a soft kernel. Furthermore, this approach may be used for segmentation pre-processing, or visual comparison of CT volume renders [9].

Buades, Coll, and Morel, described a patch-based non-local-means technique for image denoising. First, the authors introduced the non-local-means approach to denoising, which considers similarities across image patches, and proved that it has significant advantages over existing methods. Next, the presented technique was detailed, including a patch-based approach for estimating denoised pixel values. Finally, the authors ended their study by underlining the importance of non-local-means denoising and its potential for further advancement [10].

Another algorithm for denoising based on Soft-thresholding, a kind of wavelet shrinking, has been used for signal denoising. The soft-thresholding process entails converting the signal to a wavelet basis and performing a non-linear transformation to reduce small wavelet coefficients to zero. This approach may denoise signals by thresholding of noisy wavelet coefficients to minimize noise, while keeping significant signal properties [11, 12].

Tanabe, N. et al. applied a deep learning algorithm to transform chest CT images generated using different methods into a uniform format, increasing the accuracy of lung nodule assessments. In addition, scientists trained a neural network to map between images acquired with various kernels, and discovered that transformed images have better consistency, resulting in more reliable measures of nodule size and volume [13]. Finally, Lee et al. used a convolutional neural network (CNN) to solve their problem, showing the ability to transform a reconstructed image to a one reconstructed with a different kernel without raw data [14].

3 Background

3.1 Bilateral Filter

Bilateral filtering is a typical image denoising technique that reduces noise, while keeping crucial image elements like edges and textures. This filter works on image de-noising by smoothing each pixel in an image based on the similarity of its value to those of its surrounding pixels, while also considering the spatial distance between them. Each pixel is subjected to a weighted average, with the weights dependent on both color similarity and spatial distance between nearby pixels. This enables the filter to smooth the image selectively, while keeping key details. The functionality of the bilateral filter for image denoising is determined by the filter parameters used, such as the spatial and range standard deviations and the window size, as shown in Fig. 1. The application and properties of the image being denoised determine the best parameter values [15]. Bilateral filtering has gained popularity and is currently employed in image processing applications such as denoising, image enhancement, and so on [16].

The bilateral filter is a modification of the Gaussian filter, because the Gaussian filter performs over-smoothing and blurring of the image, as we use the same filter for all pixels regardless of the context around each pixel.

The Gaussian filter can be expressed as:

$$g[i, j] = \frac{1}{W_s} \sum_m \sum_n f(m, n)\sigma_s[i - m, j - n] \tag{1}$$

where $\frac{1}{W_s}$ is the normalization factor, and $\sigma_s[i - m, j - n]$ is the spatial Gaussian. We need to avoid blurring with the Gaussian filter. So, we added another term to adjust the pixel brightness.

$$g[i, j] = \frac{1}{W_{sb}} \sum_m \sum_n f(m, n)\sigma_s[i - m, j - n]\sigma_b(f[m, n] - f[i, j]) \tag{2}$$

$\sigma_b(f[m, n] - f[i, j])$. is called the brightness Gaussian, while $\frac{1}{W_{sb}}$ is the normalization factor. The normalization factor is used to keep the filter energy equal to 1.

$$\frac{1}{W_{sb}} = \sigma_s[i - m, j - n]\sigma_b(f[m, n] - f[i, j]) \tag{3}$$

The advantage of the bilateral filter is that it is simple and clear to express. A weighted mean of its neighbors replaces each pixel. It depends on two parameters, which define the size and contrast of the characteristics to be preserved. Furthermore, the filter has a non-iterative procedure. This makes the settings simple to adjust, because their influence does not accumulate across several iterations [17].

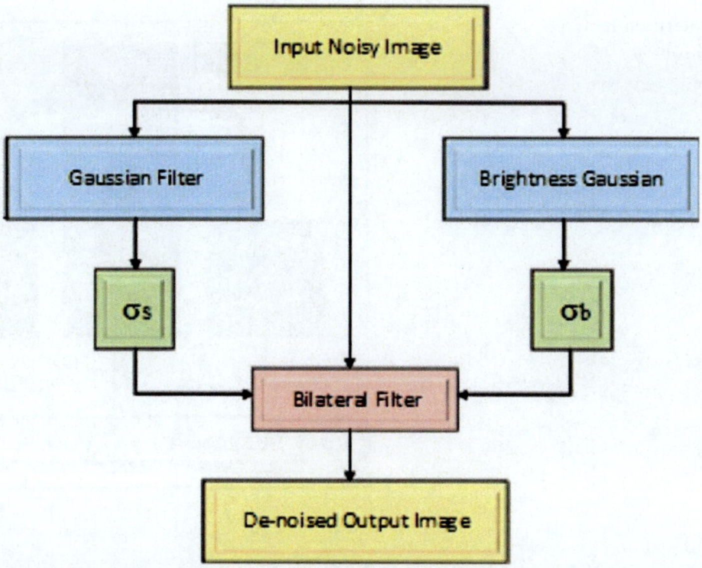

Fig. 1 Framework of Bilateral Filter.

3.2 *Quaternion Mathematics*

Quaternion is a non-commutative extension of complex numbers composed of four variables [18]. A quaternion variable is symbolized as q, which has a real part Rq that is indicated by a, and the vector vq, which consists of three imaginary parts symbolized as (b,c,d). Therefore, the q can be expressed as:

$$q = R\{q\}\ ,\ v\{q\} = [q_a\ ,\ \ (q_b, q_c, q_d)] \tag{4}$$

Where q_a, q_b, q_c and $q_d \in R$, and i, j, k are imaginary units. q_a is called the real part, and $(q - q_a)$ is the vector part (imaginary part).

According to Hamilton's relation, the multiplication between the three imaginary numbers will be as follows:

$$i^2 = j^2 = k^2 = ijk = -1, ij = -ji = k, ki = -ik = j, jk = -kj = i. \tag{5}$$

Consequently, the multiplication process cannot be commutative due to the relation between the imaginary parts. For further mathematical operation details, see [19, 20].

A color image in the RGB domain comprises three color channels: red, green, and blue. Therefore, a pure quaternion can represent a color pixel, as shown in Fig. 2:

$$quat_{image} = 0 + R\,i + G\,j + B\,k \tag{6}$$

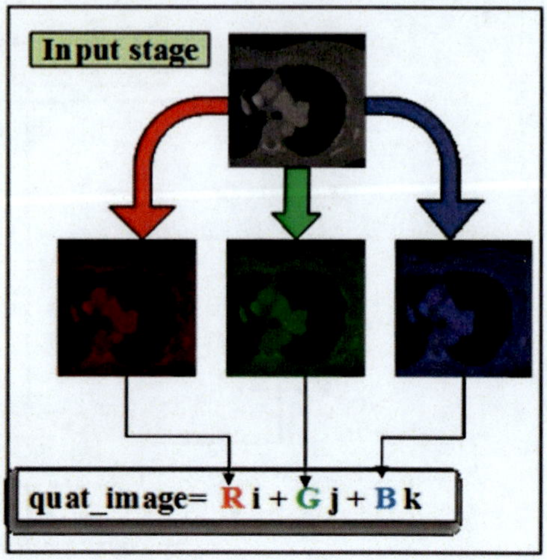

Fig. 2 Quaternion Image Representation.

The red, green, and blue pixel values are R, G, and B. As a result, a three-dimensional pixel as a whole may be processed by applying different denoising algorithms using quaternion algebra.

4 Quaternion Bilateral Filter (QBF)

The bilateral filter is a spatial-domain filter frequently used in image processing. It works on spatially-adjacent pixels in an image. The bilateral filter is based on the spatial-domain pixel intensity similarities within a small neighborhood. It applies a weighting function based on the intensity differences and the spatial distances between nearby pixels.

In the last section, we introduced quaternion mathematics and the bilateral filter. Now, we will present the proposed technique for using the bilateral filter based on quaternion mathematics for image de-noising to convert the reconstructed images with sharp kernels to ones that seem to be reconstructed with soft kernels without losing the pivotal data from the image.

Firstly, we represent our input image in a quaternion form as shown in Eq. (6). We convert every image pixel into a quaternion (vector), and then instead of working on a single value, we work with a vector for every pixel. Lastly, we apply the bilateral filter algorithm shown in Eq. (2) using quaternion mathematics [21] to get a quaternion bilateral filter (QBF) output image. This filter avoids over-smoothing and blurring problems, as shown in Fig. 3. Applying quaternion mathematics for each pixel gives an image with reduced noise. At the same time, the edges are preserved.

Implementation and simulation of the QBF were performed with Matlab code. This filter may be used on both grayscale and color images. The filter is initially

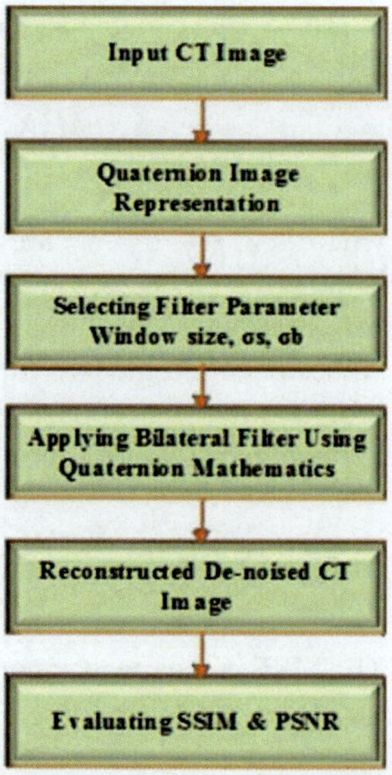

Fig. 3 Flowchart for the proposed QBF (Bilateral Filter Using Quaternion Mathematics).

applied on the input image inside the window of choice. The image becomes more blurry, when the σ_s is increased. The window size is then modified, and the range filter is computed. Lastly, before normalizing the output, the bilateral filter is generated by taking the product of the domain and range filters using the quaternion product [19, 20].

5 Experimental Results

5.1 Evaluation Metrics

After QBF implementation for image denoising, we have to compare the output from the sharp kernel input to that of the soft kernel one. So we have two factors for comparison and evaluation: a) visual analysis and b) metric analysis. For performing visual analysis, there are no mathematical or numerical tasks to be done. It depends on the visibility and clarity of the image quality and its details [2]. Then, to measure the performance, different metrics can be used for analysis. This paper depends on the PSNR and SSIM as performance metriccs. All images used are CT images with the size of 512×512 and 16 bits in DICOM format. They are generated from the raw data using different kernels. Each kernel works on 13 different CT-scanned images. So, our experiment results depend on calculating the average SSIM and PSNR for the 13 different images. Our experiments were done using MATLAB.

The SSIM can be used to compare two input images, X and R. It can be calculated as,

$$SSIM(X, R) = \frac{(2\mu_X\mu_R + C_1)(2\sigma_{XR} + C_2)}{(\mu_X^2 + \mu_R^2 + C_1)(\sigma_X^2 + \sigma_R^2 + C_2)} \tag{7}$$

where μ_X, μ_R, σ_X^2, σ_R^2 are averages and variances of X and R, respectively, and $\sigma_X R$ is the covariance between X and R. C_1 and C_2 are constant to avoid the division-by-zero problem. The SSIM shows the accuracy of the structure and similarity of pixel values between the two input images [22].

The other metric used is PSNR, and it is an essential metric for image denoising performance assessment. It can be calculated as,

$$PSNR = 10 \times \log_{10}\left(\frac{255 \times 255}{MSE}\right) \tag{8}$$

The MSE is the mean square error between the original and the reconstructed images,

$$MSE = \frac{1}{mn}\sum_{i=0}^{m-1}\sum_{j=0}^{n-1}[X(i, j) - R(i, j)]^2 \tag{9}$$

The higher PSNR value indicates a good reconstructed image. Still, the higher value does not imply that the image visual quality is good. So, the denoising does not only depend on the PSNR value [22].

5.2 Results

Our proposed QBF is applied for different kernels on CT images. At first, we do our test to find the best parameters, σ_s, σ_b as shown in Fig. 5. Then, we can see that by increasing the value of σ_s, the image gets blurred, and by changing the value of σ_b, we control the brightness and luminance of the image pixels as shown in Fig. 4. Then, we use the best value of $\sigma_s = 1$ to avoid image blurring. We check the parameter space for QBF, NLM filtering, and Wavelet image denoising as shown in Fig. 5. Samples of the cross-sections reconstructed with the B50f kernel and filtered with the used filters are shown Fig. 6. .

After adjusting the parameters, we measured the SSIM of the reconstructed images. By comparing the original B50f reconstructed image before and after applying the filters with the references reconstructed with B30f using a 5×5 window size, We found that the SSIM for the original B50f compared to the reference images is 0.84648. In contrast, the SSIM result between the B50 reconstructed image using the proposed QBF is 0.950088. By comparing the proposed technique to different denoising techniques, such as NLM filtering [10] and wavelet denoising [12], the results are 0.940356 and 0.865829, respectively, for B50f outputs used as input to

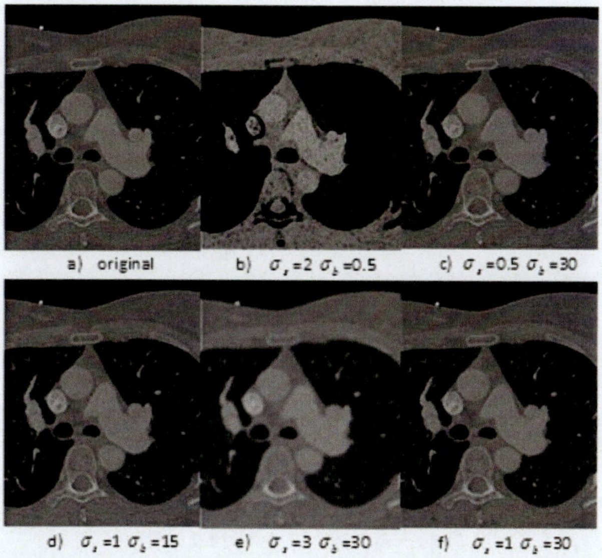

Fig. 4 The cross-sections were reconstructed with the B50f after applying the QBF with different parameters.

Table 1 Average SSIM values for different kernels and filters with B30f reference.

Kernel	Original Images	QBF	NLM	Wavelet Denoising
B50f	0.84648	0.950088	0.940356	0.865829
B46f	0.9648	0.968455	0.959793	0.96144
B41f	0.98079	0.959438	0.970493	0.972753
B36f	0.98127	0.956093	0.971545	0.971441
B35f	0.98791	0.953844	0.974319	0.974208
B31f	0.98555	0.962748	0.973425	0.97503
Mean	0.9578	0.958444	0.964989	0.95345
St. Dev.	0.00356	0.009341	0.002475	0.003285

these filters. Finally, the average SSIM and PSNR results were compared to different images reconstructed with different kernels after applying other denoising techniques as shown in Table 1. The average PSNR values between the original and reconstructed images are listed in Table 2. From these results, we can see that the proposed technique is not a satisfactory in terms of PSNR improvement compared to other techniques. Still, the high PSNR does not mean that the visual quality of the reconstructed or denoised image is good [7].

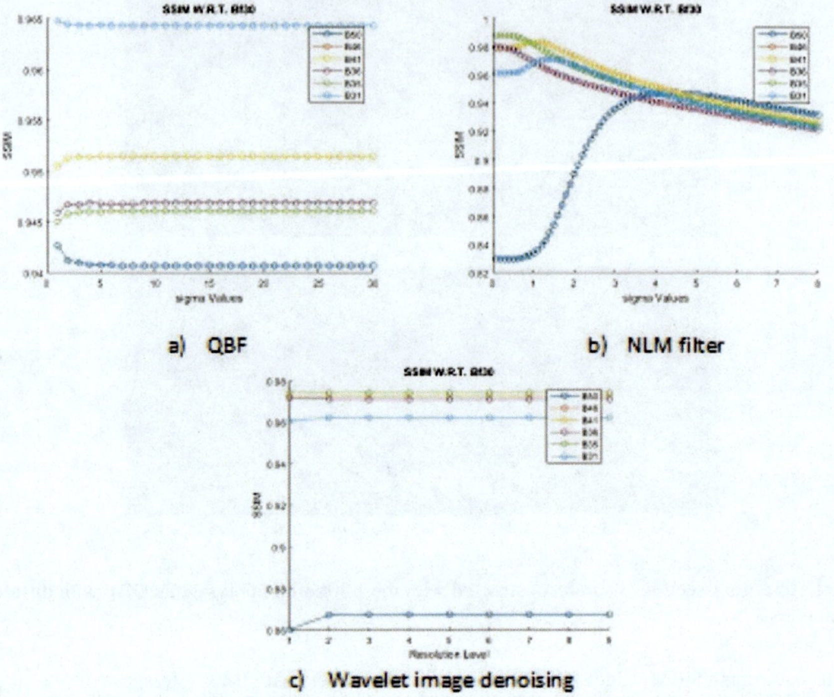

Fig. 5 SSIM values for (a) QBF. (b) NLM filter. (c) Wavelet image denoising.

Table 2 Average PSNR values for different kernels and filters with the original image.

Kernel	QBF	NLM	Wavelet Denoising
B50f	39.99023	38.20174	43.14137
B46f	33.31209	46.18957	46.2211
B41f	34.21419	47.35876	47.15281
B36f	33.2505	48.94576	48.7765
B35f	34.96675	49.16635	49.00006
B31f	36.57453	47.9977	47.7785

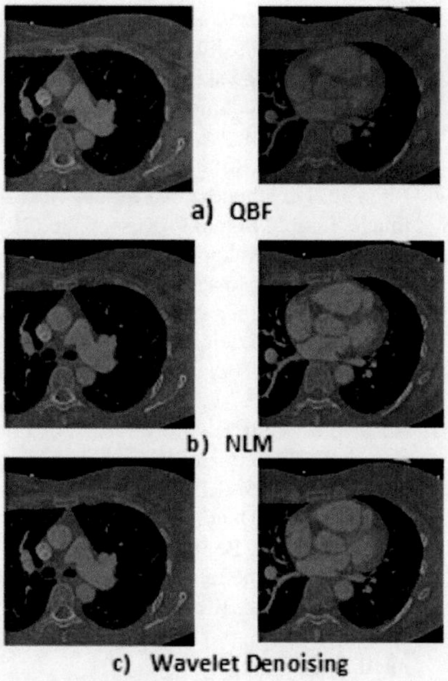

Fig. 6 One of the cross-sections reconstructed with the B50f and filtered with QBF (a) reconstructed with the B50f and filtered with the NLM filter (b) reconstructed with the B50f and filtered with Wavelet denoising (c).

6 Conclusion

To denoise medical images, we presented the suggested QBF. We implemented solutions such as non-local means filtering and Wavelet denoising for comparison. The proposed QMF has a good performance. It removes noise from B50f and B46f outputs. However, the disadvantage of this filter is that it cannot remove noise efficiently with medium kernels. It has a non-iterative procedure. Furthermore, it is not free of parameters. With no access to raw data, the QBF can efficiently produce images with less noise. The performance of the proposed QBF is determined by the window size (W) and standard deviations (σ_s , σ_b).

References

1. Mohd Sagheer, S.V., George, S.N.: A review on medical image denoising algorithms. Biomed. Signal Process. Control **61**, 102036 (2020)
2. Diwakar, M., Kumar, M.: A review on CT image noise and its denoising. Biomed. Signal Process. Control **42**, 73–88 (2018)
3. Hounsfield, G.N.: Computerized transverse axial scanning (tomography): Part 1. Description of system. Br. J. Radiol. **46**, 1016–1022 (1973)
4. Güler, E., et al.: Effect of iterative reconstruction on image quality in evaluating patients with coronary calcifications or stents during coronary computed tomography angiography: a pilot study. Anatol. J. Cardiol. **16**, 119 (2016)

5. Mileto, A., Guimaraes, L.S., McCollough, C.H., Fletcher, J.G., Yu, L.: State of the art in abdominal CT: the limits of iterative reconstruction algorithms. Radiology **293**, 491–503 (2019)
6. Lasek, J., Piórkowski, A.: CT scan transformation from a sharp to a soft reconstruction kernel using filtering techniques. In: Singh, S.K., Roy, P., Raman, B., Nagabhushan, P. (eds.) CVIP 2020. CCIS, vol. 1376, pp. 56–65. Springer, Singapore (2021). https://doi.org/10.1007/978-981-16-1086-8_6
7. Bhonsle, D., Chandra, V., Sinha, G.R.: Medical image denoising using bilateral filter. Int. J. Image, Graph. Signal Process. **4**, 36 (2012)
8. Kumar, M., Diwakar, M.: A new exponentially directional weighted function based CT image denoising using total variation. J. King Saud Univ. Inf. Sci. **31**, 113–124 (2019)
9. Kikinis, R., Pieper, S.D., Vosburgh, K.G.: 3D slicer: a platform for subject-specific image analysis, visualization, and clinical support. In: Jolesz, F.A. (ed.) Intraoperative Imaging and Image-Guided Therapy, pp. 277–289. Springer, New York (2014). https://doi.org/10.1007/978-1-4614-7657-3_19
10. Buades, A., Coll, B., Morel, J.-M.: A non-local algorithm for image denoising. In: 2005 IEEE Computer Society Conference on Computer Vision and Pattern Recognition (CVPR 2005), vol. 2, pp. 60-65. IEEE (2005)
11. Donoho, D.L.: Denoising by soft-thresholding. IEEE Trans. Inf. Theory **41**, 613–627 (1995)
12. Abramovich, F., Benjamini, Y., Donoho, D.L. Johnstone, I.M.: Adapting to unknown sparsity by controlling the false discovery rate (2006)
13. Tanabe, N., et al.: Kernel conversion for robust quantitative measurements of archived chest computed tomography using deep learning-based image-to-image translation. Front. Artif. Intell. **4**, 209 (2022)
14. Lee, S.M., et al.: CT image conversion among different reconstruction kernels without a sinogram by using a convolutional neural network. Korean J. Radiol. **20**, 295–303 (2019)
15. Zhang, M., Gunturk, B.K.: Multiresolution bilateral filtering for image denoising. IEEE Trans. Image Process. **17**, 2324–2333 (2008)
16. Tomasi, C., Manduchi, R.: Bilateral filtering for gray and color images. In: Sixth International Conference on Computer Vision, pp. 839–846. IEEE (1998). (IEEE Cat. No. 98CH36271)
17. Liao, Z., Hu, S., Yu, Z., Sun, D.: Medical image blind denoising using context bilateral filter. In: 2010 International Conference of Medical Image Analysis and Clinical Application, pp. 12-17. IEEE (2010)
18. Hamilton, W.R.: Scientific books: elements of quaternions. Science **14**, 65–66 (1901)
19. Hamilton, W.R.: Elements of Quaternions. Longmans Green, & Company, Harlow (1866)
20. Zhang, F.: Quaternions and matrices of quaternions. Linear Algebra Appl. **251**, 21–57 (1997)
21. Lewis, D.W.: Quaternion algebras and the algebraic legacy of Hamilton's quaternions. Irish Math. Soc. Bull. **57**, 41–64 (2006)
22. Sagheer, S.V.M., George, S.N.: A review on medical image denoising algorithms. Biomed. Signal Process. Control **61**, 102036 (2020)

Exploring Text-Guided Synthetic Distribution Shifts for Robust Image Classification

Ryan Ramos, Raphael Alampay, and Patricia Abu

Abstract The empirical risk minimization approach of contemporary machine learning leads to potential failures under distribution shifts. While out-of-distribution data can be used to probe for robustness issues, collecting this at scale in the wild can be difficult given its nature. We propose a novel method to generate this data using pre-trained foundation models. We train a language model to generate class-conditioned image captions that minimize their cosine similarity with that of corresponding class images from the original distribution. We then use these captions to synthesize new images with off-the-shelf text-to-image generative models. We show our method's ability to generate samples from shifted distributions, and the quality of the data for both robustness testing and as additional training data to improve generalization.

Keywords Robustness to distribution shift · synthetic data · foundation models

1 Introduction

Contemporary machine learning models are heavily reliant on data, to the point that at smaller scales of training data, models fail to adapt to novel datapoints outside their original training distribution [6]. A classic illustration of this is image classification: if images of cows in a model's training distribution frequently involved grassy fields, the said model is likely to fail in identifying cows when they are present in beaches [3]. This phenomenon is a consequence of combining the empirical risk minimization paradigm and a non-large-scale data regime. With a limited training distribution, there is less information available to models for them to disentangle spurious correlations from target labels, and thus these shortcuts become leveraged as predictive features despite their lack of robustness under distribution shifts [6, 20].

To address this issue in practice, researchers evaluate their models on test sets in order to examine how well a model adapts to data it has explicitly not been optimized

R. Ramos (✉) · R. Alampay · P. Abu
ALIVE, Ateneo de Manila University, Quezon City, Philippines
e-mail: ryan.ramos@obf.ateneo.edu

R. Burduk et al. (eds.), *Progress on Pattern Classification, Image Processing and Communications*, Lecture Notes in Networks and Systems 766,
https://doi.org/10.1007/978-3-031-41630-9_16

for during training. However, these test sets tend to be independent and identically distributed, and potentially share the same underlying distribution as the training data [20], thus eliminating the effectiveness of using a test set to probe the ability of a model to adapt to new domains. In line with this, studies have taken steps towards building specifically-engineered datasets and benchmarks to probe models' abilities to generalize to out-of-distribution data [8, 13].

While many traditional datasets are carefully sourced and manually annotated, recent advancements in generative modelling and foundation models have proven able to generate clean, usable data at scale without being restricted to natural data [1, 11, 23]. More relevant to this study, they have also been used to successfully simulate distribution shifts in data [12, 22].

A persistent question in the field of synthetic distribution shifts is how far of a distribution shift can we successfully simulate, and what methods make synthetic distribution shifts possible. We continue this line of work by proposing a novel method for this task. We train a language model to generate captions that maximize their CLIP distance from reference classes, and use these captions to steer generative image models towards outputs that align with predefined classes but do not represent the original distribution. Our method generates data that is on average 15% lower in CLIP similarity to images from the same class when compared to in-distribution data. As a robustness benchmark, our method reduces classification accuracy by approximately 40%. Our generated data also leads to performance gains when leveraged as additional training data.

2 Related Work

Given a distribution of training data, an out-of-distribution dataset would introduce some distribution shift relative to the original distribution. Specifically, if the training distribution is said to be the result of some causal model or environment E, the new dataset should be sampled from E', the product of intervening on or changing the joint distribution of E. When evaluated on a new joint distribution, model A is said to be more robust than model B if it sees a smaller drop in performance resulting from leveraging reliable features that are invariant across domains; less robust models would be exposed for relying on spurious correlations that are not invariant across domains.

Probing for robustness is traditionally performed with curated datasets, many of which require manual annotation or image post-processing. With ImageNet assumed as a train set for the task of image classification, distribution shifts include varying viewpoints [2], rendering images in different styles [8, 23], corrupting or perturbing the images [9], etc. Another popular approach would be to specifically engineer an adversarial benchmark, where data is selected particularly based on a reference model's inability to correctly classify the data with high confidence [10].

As curated datasets, these methods are essentially limited by what data is naturally available. Generative models have provided ways to artificially induce distribution shifts in data, such as texture changes [15] and optimally adversarial augmentations [7]. Particularly relevant to this study are methods such as [22] and D3S [12], which leverage recent advancements in diffusion models [4, 18] to generate out-of-distribution data for image classification. While these works either utilize simple yet effective prompting techniques with frozen models, or leverage gradient information in the diffusion process, our method investigates a new space for optimizing generative model outputs for out-of-distribution data. We specifically investigate optimizing not the diffusion process, but a text generation process prior to text-guided diffusion instead. We discuss this further in Sect. 3.

3 Distribution-Shifted Data Generation

Given a distribution of n images $I_C := \{I_0, I_1, \ldots, I_n\}$ sampled from some class C, our proposed method generates images from that class that fall under C but lie away from the original distribution. We achieve this by leveraging a pretrained text-to-image model G. Unlike previous works [12, 22], we leave the mechanics of G untouched but optimize the text generation process that would steer G.

In order for the generated images to lie away from the original distribution, the corresponding texts that will guide the image generation must be sensitive to the original data and not represent its distribution. To implement this, we consider the goal of maximizing the distance of our generated captions, and consequently generated images, from the original distribution in a shared image-text embedding space modelled by some multimodal embedder $\phi(\cdot)$. Thus, we finetune a generative language model L to generate a caption P_C conditioned on a given class C that minimizes the following objective function R:

$$R(P_C) = s\left(\phi(P_C), \frac{\sum_{i=0}^{n} \phi(I_i)}{n}\right) \tag{1}$$

where $s(\cdot, \cdot)$ measures the similarity between two points in an embedding space, with higher values indicating stronger similarity. This value compares a generated caption with the mean of ϕ-encoded images of I_C, thus minimizing this value would maximize the distance of the caption from the original data distribution. Because a language model can create arbitrary noise to maximize the distance of its outputs from a reference distribution in the embedding space, we further modify R by returning the lower bound of s for any any non-sensical (incomplete sentence and failure to mention class C in prompt) output. Generated captions can then be used as prompts for G without further finetuning or modifications.

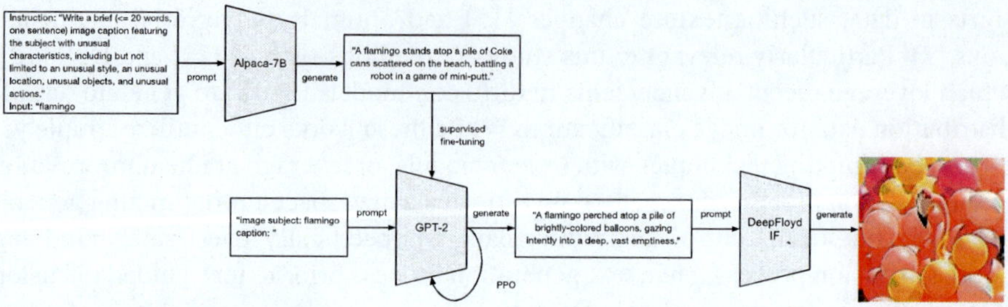

Fig. 1 Overview of method.

4 Experiments

4.1 Implementation

For our experiments, we focus on the distribution of images found in ImageNet. Specifically, we focus on the 86 class susbset intersection between ImageNet-A [10] and ImageNet-R [8] for comparability purposes. To calculate mean embeddings for each class, we sample 128 images per class.

An overview of the generation pipeline is shown in Fig. 1. We leverage DeepFloyd IF, an open-access implementation of Imagen [19], for G; CLIP [16] for ϕ; and base our L on GPT-2. During inference, we use the generative image model at 16-bit precision, with its text encoder components in 8-bit precision; and the latter two models at 8-bit precision. Following a two-stage generate-upsample pipeline, we sample using 27 steps and upscale with 10 steps. Our similarity scorer s is cosine similarity. We also use CLIP cosine similarity as a filter to remove low-quality data from our generations, using a threshold of 0.2.

In training L, we first use an Alpaca-7B to generate 16 sample captions per ImageNet class to create 16000 examples of class-conditioned prompts. We then use this data to perform supervised finetuning on a GPT-2 followed by reinforcement learning with PPO [21].

For all reference methods, in order to maintain comparability as regards image fidelity, we reimplement each method using the same generative image model backbone, model precision, and diffusion sampling steps.

Sample generations of our method are presented in Fig. 2.

4.2 Measuring Distribution Shift

Table 1 describes our generated data compared to both the original data and similar methods. Mahalanobis distance was calculated in CLIP space, reduced to 64 dimensions via PCA. While our method does not outperform other methods in CLIP

Fig. 2 Sample generated texts with corresponding generated images for the ImageNet class "school bus".

Table 1 Distance metrics describing relation of datapoints to class mean.

Method	CLIP similarity ↓	Mahalanobis distance ↑
Base distribution	83.50	7.90
Sehwag et al. 2022 [22]	70.26	24.70
D3S [12]	73.74	24.49
Ours	73.93	23.30

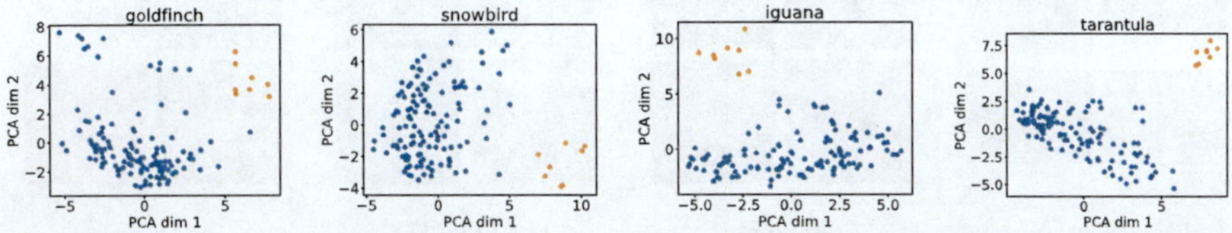

Fig. 3 PCA-reduced plots of datapoints in CLIP space. Blue points represent the original distribution while orange points represent our generated data. (colour figure online)

similarity or Mahalanobis distance, we maintain similar performance as regards maintaining a gap from the base distribution. Examples of this gap can be visualized in Fig. 3, which shows examples in a PCA-reduced space. Furthermore, Fig. 4 shows the percent reduction in cosine similarity of datapoints to the class mean in CLIP space across all examined classes, with our method only creating datapoints closer to the mean in one class while reducing similarity by approximately 9.5% on average.

4.3 Probing Adversarial Ability

We also measure the strength of our distribution shift via zero-shot evaluation on our data, where pretrained classifiers are expected to drop in accuracy under stronger

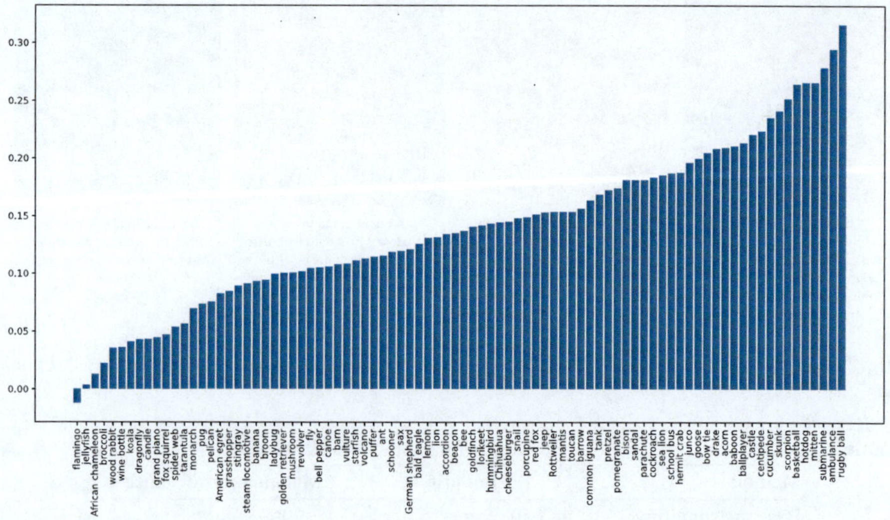

Fig. 4 Percent reduction in average cosine similarity of generated datapoints in a class compared to original datapoints from the same class, in CLIP space.

(a) ImageNet (b) Low-density regions [22] (c) D3S [12] (d) Ours

Fig. 5 Images of the ImageNet class "German shepherd" drawn from (5a) the ImageNet training set, (5b) sampling from low-density regions, (5c) generating images with different backgrounds, and (5d) our proposed method of generating captions to guide out-of-distribution image generation.

Table 2 Accuracy reduction of ImageNet classifiers on distribution shift validation datasets.

Method	ConvNeXt-Atto [14]	ViT-Tiny [5]
Low-density [22]	−62.04	−58.73
D3S [12]	−34.71	−43.04
Ours	−39.51	−41.87

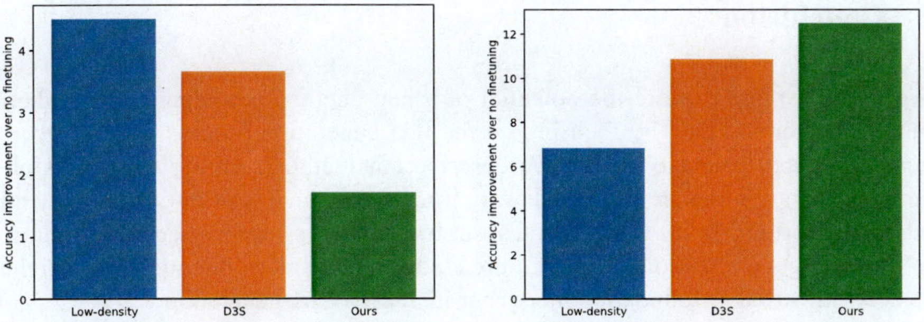

(a) ImageNet-A [10] (b) ImageNet-R [8]

Fig. 6 Accuracy improvements on adversarial datasets by fine-tuning on distribution shift datasets.

distribution shifts, assuming the model relies on spurious correlations rather than causal features. The results of these experiments are seen in Table 2. Though we do not outperform both methods, our method is able to reduce classification accuracy by around 40% on each model, indicating the potential for this method to act as a challenging benchmark for image models.

4.4 *Use as Training Data*

Aside from acting as a robustness benchmark, another potential use of out-of-distribution data is to act as training data in order to widen the training distribution and induce better generalization ability during inference. We fine-tune a ConvNeXt-Atto [14] on data generated with our method as well as two other synthetic distribution-shift methods, with 8 generated samples per class, then evaluate performance gains in the out-of-distribution setting, in this case ImageNet-A and ImageNet-R. Our results are seen in Fig. 6.

We show the most improvement on ImageNet-R, while introducing the least improvement on ImageNet-A. We believe this to be attributed to the data generation process of each method. While our method produces a mix of realistic and artistically rendered images, the other methods mainly produce photo-like images. At the current data scale, the other methods have a greater impact for ImageNet-A, which mainly includes photos, while our data has the most impact for ImageNet-R, which does not exclusively focus on photos.

5 Conclusion

We introduce and explore the potential of a new method for generating synthetic out-of-distribution data by focusing on the text generation process used to steer a generative text-to-image model. We describe our method's ability to successfully generate samples that are distanced away from the original distribution, and quantitatively report its ability to act as a test set for robustness probing and as a train set to increase generalization ability. Future studies can explore this method at higher scales with stronger foundation models or increased data generation.

References

1. Azizi, S., Kornblith, S., Saharia, C., Norouzi, M., Fleet, D.J.: Synthetic data from diffusion models improves imagenet classification. arXiv preprint arXiv:2304.08466 (2023)
2. Barbu, A., et al.: ObjectNet: a large-scale bias-controlled dataset for pushing the limits of object recognition models. In: Advances in Neural Information Processing Systems, vol. 32 (2019)
3. Beery, S., Van Horn, G., Perona, P.: Recognition in terra incognita. In: Proceedings of the European conference on computer vision (ECCV), pp. 456–473 (2018)
4. Dhariwal, P., Nichol, A.: Diffusion models beat Gans on image synthesis. Adv. Neural. Inf. Process. Syst. **34**, 8780–8794 (2021)
5. Dosovitskiy, A., et al.: An image is worth 16x16 words: transformers for image recognition at scale. arXiv preprint arXiv:2010.11929 (2020)
6. Geirhos, R., et al.: Shortcut learning in deep neural networks. Nat. Mach. Intell. **2**(11), 665–673 (2020)
7. Geirhos, R., Rubisch, P., Michaelis, C., Bethge, M., Wichmann, F.A., Brendel, W.: ImageNet-trained CNNs are biased towards texture; increasing shape bias improves accuracy and robustness. arXiv preprint arXiv:1811.12231 (2018)
8. Hendrycks, D., et al.: The many faces of robustness: a critical analysis of out-of-distribution generalization. In: Proceedings of the IEEE/CVF International Conference on Computer Vision, pp. 8340–8349 (2021)
9. Hendrycks, D., Dietterich, T.: Benchmarking neural network robustness to common corruptions and perturbations. arXiv preprint arXiv:1903.12261 (2019)
10. Hendrycks, D., Zhao, K., Basart, S., Steinhardt, J., Song, D.: Natural adversarial examples. In: Proceedings of the IEEE/CVF Conference on Computer Vision and Pattern Recognition, pp. 15,262–15,271 (2021)
11. Honovich, O., Scialom, T., Levy, O., Schick, T.: Unnatural instructions: Tuning language models with (almost) no human labor. arXiv preprint arXiv:2212.09689 (2022)
12. Kattakinda, P., Levine, A., Feizi, S.: Invariant learning via diffusion dreamed distribution shifts. arXiv preprint arXiv:2211.10370 (2022)
13. Koh, P.W., et al.: Wilds: a benchmark of in-the-wild distribution shifts 2021. arXiv preprint arXiv:2012.07421 (2020)
14. Liu, Z., Mao, H., Wu, C.Y., Feichtenhofer, C., Darrell, T., Xie, S.: A convnet for the 2020s. In: Proceedings of the IEEE/CVF Conference on Computer Vision and Pattern Recognition, pp. 11,976–11,986 (2022)
15. Pintor, M., et al.: ImageNet-patch: a dataset for benchmarking machine learning robustness against adversarial patches. Pattern Recogn. **134**, 109,064 (2023)
16. Radford, A., et al.: Learning transferable visual models from natural language supervision. In: International Conference on Machine Learning, pp. 8748–8763. PMLR (2021)

Artificial Intelligence: Various Applications and Problems

Color Tracking Application Using AI-Based Docker Container Scheduling in Fog Computing

Rafal Chrobak, Sebastián García Galán, Enrique Munoz Expósito, Manuel Valverde Ibanez, Tomasz Marciniak, and Adam Marchewka

Abstract This paper presents the real implementation of a fog computing environment for the execution of color tracking applications by using FogBus2 framework and an artificial intelligence based docker container scheduling. To be precise, an edge computing network has been developed by using a personal computer and several small computing devices such as Raspberry Pi and Nvidia Jetson Nano. Related to the scheduling policy, besides the existing policies in Fogbus2 framework, another one based on fuzzy rules-based system has been designed. Results demonstrate the proposed policy outperforms classical approaches, even when using, pavin the way to the use of knowledge acquisition techniques in order to improve the scheduling performance in terms of makespan and flowtime.

1 Introduction

Due to Internet of things (IoT), computing paradigms have been evolving through decentralized environments. Thus, cloud computing was initially used since it provided computing infrastructures, allowing de the development of new services on the internet [1, 2]. However, IoT considers an overwhelming number of smart devices with new requirements in terms of latency, for instance, given rise to other computing paradigms to face new challenges around IoT revolution.

In this regard, edge computing emerged as a computing paradigm with the capability of dealing with those hindrances cloud computing could not face appropriately. In fact, edge computing, by means of several different implementations, provides users with some advantages such as:

R. Chrobak · T. Marciniak · A. Marchewka
Faculty of Telecommunications, Computer Science and Electrical Engineering, Bydgoszcz University of Science and Technology, Bydgoszcz, Poland

S. G. Galán (✉) · E. M. Expósito · M. V. Ibanez
Telecommunications Engineering Department, University of Jaén, Linares (Jaén), Spain
e-mail: sgalan@ujaen.es

R. Burduk et al. (eds.), *Progress on Pattern Classification, Image Processing and Communications*, Lecture Notes in Networks and Systems 766,
https://doi.org/10.1007/978-3-031-41630-9_17

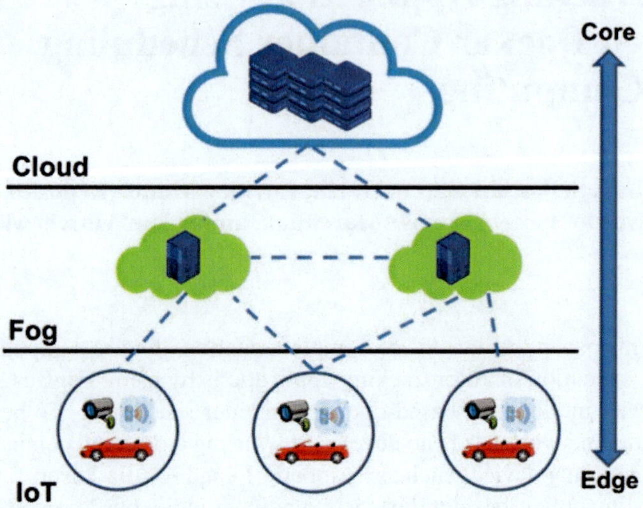

Fig. 1 Fog Computing Architecture.

- Better bandwidth usage (compared with cloud computing) due to the transmission data collected from many sensors in only one packet.
- Better geographical distribution for devices since not all of them need internet connection.
- Sensors can be simplest devices because of the use of local connection with edge router. Thus, smaller, cheaper and more efficient devices in terms om energy consumption could be used.

In this context, fog computing, whose general three-layer architecture is depicted Fig. 1, represents the highest evolution the edge computing principles.

As it can be observer in Fig. 1, there are three layers called Cloud, Fog and IoT [1]. Cloud Layer represents the more centralized architecture, composed by servers with high computational capability and enough storage resources. On its own, fog layer represents the core of the fog layer architecture composed by physical and logical network elements implementing fog computing services [3], which could imply computation, transmission and temporary storage of data. Finally, IoT layer is composed by geographically distributed IoT devices such as sensors, among others. Therefore, fog computing implies a complete architecture of distributed resources along by means of an extension of cloud computing services. Moreover, since fog computing is at the edge of the network, it provides low latency, which improves quality of services (QoS) for real-time applications. There is an increasing number of applications based on fog computing including industrial automation, transportation, and sensor and actuator networks.

In this regard, despite the fact that there are some frameworks to manage integrated resources in edge/fog computing [4, 5], they barely consider platform independent techniques, scheduling of heterogeneous IoT applications with complex dependent structures, scalability mechanisms of distributed resource managers, and containerization. In this sense, FogBus2 [6] is a lightweight and distributed container-based Framework for integration of IoT-enabled systems with heterogeneous multi-cloud and multi-edge resources. FobgBus2 offers a scalability mechanism to ensure efficient responsiveness when either the number of IoT devices increases or the resources become overburdened.

To check the scheduling capabilities of Fogbus2, this paper presents a color tacking application for its use in fog computing environments by using an intelligent approach for the scheduling of docker containers. To be precise, this work considers, as scheduling policy, the use of an expert systems through by means if fuzzy rules-based systems (FRBS) [7].

The remainder of the paper is organized as follows. Section 2 presents a background regarding the main ideas related to FogBus2. Interactions among different actions deserves special mention; therefore, they are presented in Sect. 3. The proposed solution based on FRBS is introduced in Sect. 4. Section 5 focuses on experimental results of the proposed algorithm and provides comparisons with traditional schemas. Finally, Sect. 6 concludes the paper.

2 FogBus2 Framework

In order to take advantages of new computations paradigms, FogBus2 framework was developed with an emphasis on containerization. To achieve this goal, general sub-component of this software was created as docker containers. Description in detail of this components can be found in [6, 8], and are listed below:

- User. This component is implemented in IoT device such as a sensor, a video camera, or any other device that can provide data to be used in tasks. Its main purpose is to collect data from environment and send it to another component, called Master.
- Master. This component is running in edge/fog layer, and its main purpose consists on receiving requests from users and managing the different execution tasks. Therefore, it is the component devoted to the development of the scheduling policy regarding the current capabilities of the different actors.
- Actor. This component can run on host devices in edge/fog layer and its main goal is the reception of tasks from Master and starting suitable Task Executor component.
- Task Executor. It is in charge of the execution of the different provided tasks and it is running on the same device as the Actor component, which receives tasks from Master component.
- Remote Logger. It is in charge of collection in periodic or event-driven logs from component in the network. These logs can be stored ion databases or by means of files system.

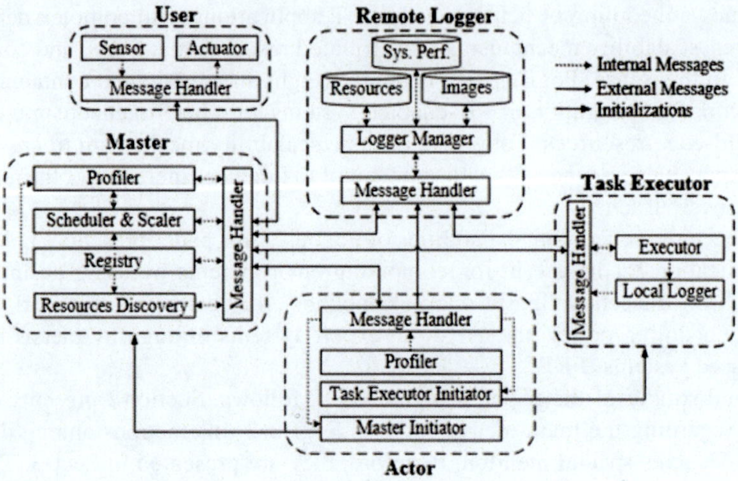

Fig. 2 Software components of Fogbus2 and their interactions.

– Data Base. In case of using database, instance of this should be running as another docker container. It can be implemented in any host device in edge/fog layer.

The interaction among these software components is depicted in Fig. 2.

All components can be deployed on different devices, but it is possible to run different containers on the same device. In addition, bearing in mind that FogBus2 framework support Virtual Private Network technology, some part of system can be implemented in Cloud using VPN. Main software components of FogBus2 framework is presented in Fig. 3.

2.1 Multiplatform Support

IoT systems and edge/fog computing networks have to be operable on many platforms. According to most popular solutions (AMD, ARM, Intel) different architectures are supported by FogBus2. This is important point to really take advantage of possibility where different devices can be implemented. This is also good solution to upgrade existing solutions with already deployed devices. To achieve this, docker images implemented in framework have the same names but during pulling images from different devices, appropriate and compatible with implemented hardware image is going to be deployed on the device.

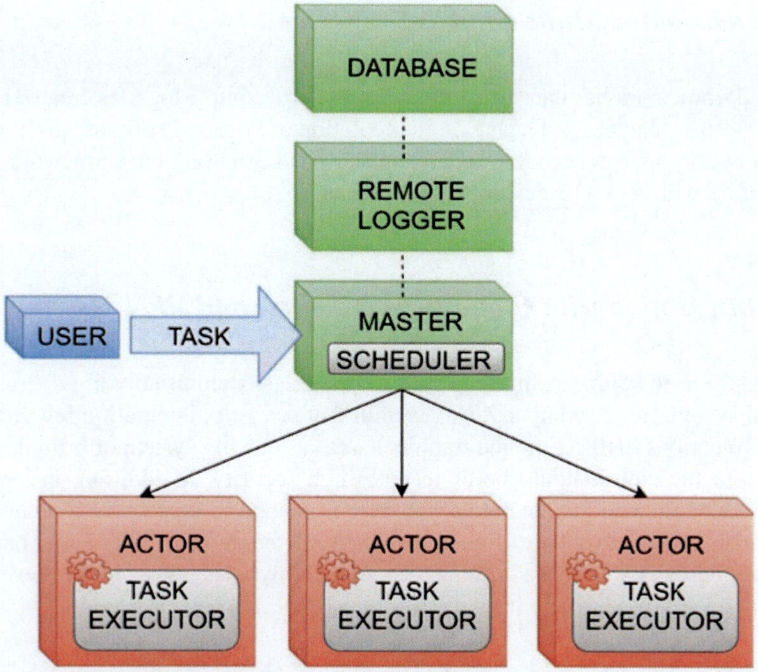

Fig. 3 Main software component of Fogbus2 framework.

2.2 Containerization Support

Main idea of Internet of Things is to create network with huge number of devices. That devices can be in charge to operate different functionality. In that scenario, big and complicated applications can be distracted to different smaller parts that they can be implemented on different devices. Implementation of FogBus2 included in this work also has taken advantage of containerization.

2.3 Heterogeneous IoT Applications

FogBus2 is environment oriented on IoT applications. Therefore, many of them can be sensitive to value of latency, especially applications that are going to operate in real-time. There is also possibility to initialize applications that needs strong computation resources. FogBus2 supports big applications that can be distributed as several number of tasks and deployed in parallel on different devices. This increasing efficiency and reduce executing time. User can use already implemented applications (like color tracking, face detection) but also modernize them or even develop completely new one application, according to needed goals.

2.4 Distributed Multi-database

Selected framework has the capability to operate with simple log files created by user or big common databases. During implementation of FogBus2 software in developed edge network, open-source MariaDB database has been used, but framework is also compatible with AutoDB developed by Oracle.

2.5 Integration with Cloud Infrastructure and VPN

Commonly used solutions implemented in modern system usually are operating as internal system inside company. But even in that scenario, integration few independent applications in different geographical areas as one big system or bringing some features to the cloud should be done with high security. In addition, in technical way (with public and private networks) there is needed to support VPN technology. FogBus2 is taking advantage of that - there is implemented P2P VPN script based on the Wireguard. This is also additional functionality what is important for developers working on solutions where high privacy or VPN technology is needed.

2.6 Dynamic Scheduling Mechanism and Policy Integration

Planning strategy (scheduler mechanism) is important especially in systems that requires high amount of resources or when number of tasks is large. Considering the information about system state, network connection and resources of the devices, there is possibility to achieve better energy consumption and/or lower execution time and many others. FogBus2 provide opportunity to run already implemented basic schedulers (Round Robin and Random Policy). But there is no problem to extend this part of framework with own scheduling policy. This is the main contribution of this work - create a FRBS-based policy that is taking advantages of resources that are available when new task is going to be executed.

3 Communications Among Components

FogBus2 framework has been designed as multi-component environment. Therefore, communications among components have to be implemented. Typical sequence of setting all network components in order to execute task provided by User to Master component was presented below in Fig. 4. In this scenario, User is sending a Request to Master with information what service have to be provided. According to this knowledge, Master component can choose appropriate Actor which contains neces-

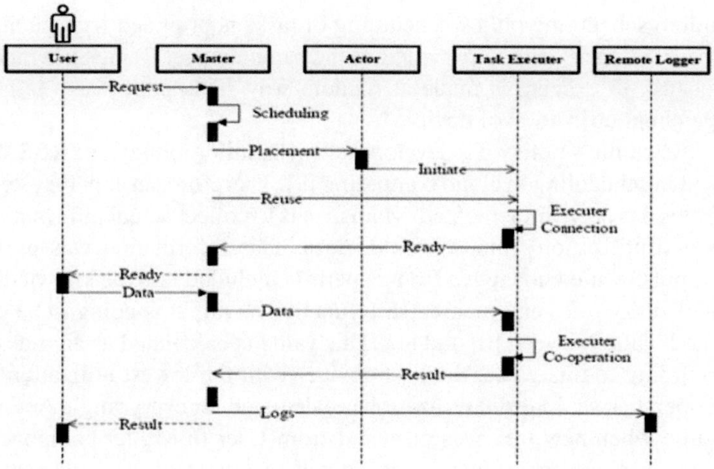

Fig. 4 Communications among Components.

sary Task Executor component. Part of Master responsibility is to plan a schedule of tasks, which is done by one of scheduling policy. After finishing schedule process, Master is providing information to selected Actor to initialize needed Task Executor. When connection is set, Task Executor component is sending "READY" statement to Master. Then Master informs User that network is ready to operate data, and User cans start sending data to initialized Task Executor. Results are sent by Task Executor to Master, and Master provides results to User and logs information are going to be deployed in Remote Logger (for example Data Base)

Interesting solution that is implemented in investigated Framework is that, after initialization of selected Task Executor, next time master can reuse existing path without request to Actor. This solution provide opportunity to save time in scheduling and setting connection among components in the future.

Container environment is an appropriate way to implement many functionalities on different devices, but there is no problem to run all containers on the same device. Communication protocol involved in this software is important part of whole system.

4 Proposed Solution for Container Scheduling in Cloud Computing

Existing policies that were used in test are Round-Robin and Random. In Round-Robin scheduling policy, scheduling of tasks is provided in circular order, and without priority. This scheduling is simple and easy to implement. It can be applied to many scheduling problems, such as data packet scheduling in computer networks. This policy is not using knowledge about utilization of devices.

In Random scheduling policy scheduling of tasks is provided without any order. In worst case, all tasks can be provided to the same device. Decision which device should get task to execute is made at random way. This policy also is not using knowledge about utilization of devices

Expert Scheduling policy To develop new scheduling policy, a FRBS has been widely used in scheduling in cloud computing [9]. Therefore, an expert system based on fuzzy rules has been implemented. Main idea is to collect actual information about resources and utilization from Actors devices. This information was provided to FRBS as inputs, while knowledge from experts is included into the knowledge bases by means of fuzzy rules and membership functions. Output is going to be provided as calculated value between [0] and [1]. This value is calculated as an index, which means according to this value, the best device (with the lowest utilization) will be chosen to operate task. Index of utilization is calculated for every single Actor device, in every time when new task was provided from User to Master component. This is the way to achieve actual information about real state of network and devices, to made appropriately decision. Inputs are collected from Actors to get information about actual state of any device. Then fuzzy system makes decisions regarding two variables

- CPU utilization
- Performance index

CPU utilization ios collected from Actor. Performance index is set up as fixed variable, for RPi it was "1" and for Jetson Nano it was "1.25". Input variables can have different values in real world - CPU utilization is from range [0, 1] and describing in what percentage resources of device are used. Performance index was set in range [0, 5].

It is important to point out that other information about resources can be obtained to get more flexible and appropriate results. In future work, creation of more complicated rules taking into account these additional variables can be done to get better information about implemented edge network. Other variables that could be considered: CPU cores, CPU frequency, Memory Max, Memory utilization and latency.

Output values can belong on one from 5 membership functions: very low, low, medium, high, very high. Shape of output membership functions was also Gaussian Function. Calculated index of utilization was value from [0] and [1] and in analyzed scenario there were no need to scale this value to another magnitude. Calculated index was used to create ranking of Actors, and Actor with higher index had a lowest utilization and was chosen to execute task.

One of the most important part of Fuzzy Rule-Based System is Rules Base. According to number of rules and that how good they describe desired decision; different output can be achieved. In this scenario, there were developed 2 different rules. This number can be higher and optimized using machine learning techniques. Used rules in this work are presented in Table 1:

In addition, an important part in developing Fuzzy System is defuzzification method. The most commonly used method in defuzzyfication process is the center of area method (COA). This is also well known as the centroid method. This

Table 1 Considered rules

INPUT CPU utilization	INPUT performance	OUTPUT
HIGH	LOW	VERY-LOW
LOW	HIGH	VERY-HIGH

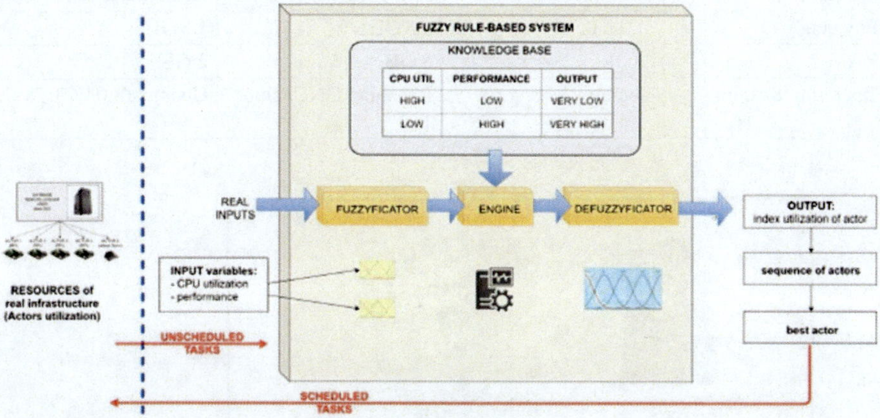

Fig. 5 Diagram of the developed Scheduling policy.

method determines the center of area of fuzzy set and as output it is returning the corresponding crisp value. This method was used in developed Fuzzy Rule-Based System. The general diagram that is describing whole system is depicted in Fig. 5.

5 Experimental Results

An important part of this work is to create real edge/fog computing network with physical infrastructure. This is testing environment for existing scheduling policies, but also for new developed schedule algorithm. Next, the used Edge computing network is going to be presented joint the experimental environment regarding a color tracking application for to test the performance of the proposed Artificial Intelligence-based scheduling policy in fog computing when using FogBus2 framework.

5.1 Developed Edge Computing Network

Some software and hardware characteristics of every used device was presented in Table 2. Diagram of developed physical infrastructure is presented in Fig. 6.

Table 2 HW and SW specification of used devices

Features	Personal Computer	Raspberry Pi 4 Model B	Nvidia Jetson Nano
Architecture	x86_64	Armv71	aarch64
Cores	4	4	4
Model	Intel Core i5-3330	ARM-Cortex-A72	ARM Cortex-A57
Frequency	3.2 GHz	1.8 GHz	1.5 GHz
RAM	8 GB	8 GB	2 GB
Operating System	Ubuntu 21.10	Raspbian GNU/Linux 11	Ubuntu 18.04.6 LTS

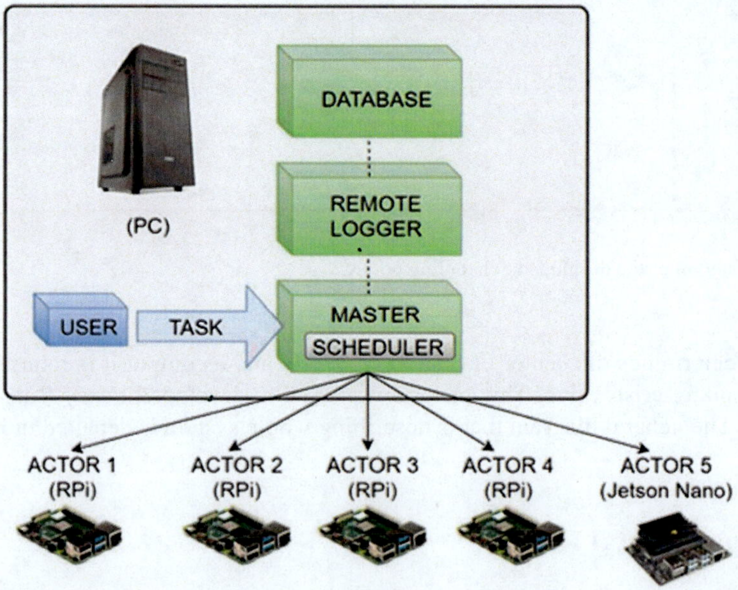

Fig. 6 Diagram of developed edge computing network.

Color Tracking application has been developed as a program in Python with GUI (Graphic User Interface). It contains 3 main windows. First is used as trackbars to set up desired values of colors that should be tracked. Next one is showing source video signal with marked colors that are actually tracked. Last one is black and white mask with marked parts of video frame wherewanted colors were find. Application is running as a program started by user component. To run it correctly, there is need to provide path to video signal that should be analyzed. It can be a video file (for example in mp4 format) but also it is possible to run application with video stream.

On Actor device has to be provided with the Task Executor component that is in charge to operate this specific task. To make whole system working appropriately,

Table 3 Videos features

Characteristic	Value
Duration time	1 s/2 s/5 s/10 s
Frames per second	15 fps
Resolution	320×240
Format of video	.mp4

this Task Executor Component has been provided to every from five involved Actors devices.

5.2 *Experimental Environment*

To check how developed scheduling policy is working, testing environment was developed. Hardware part with characteristic of used devices has been described previously. The test was to run 24 different tasks and measure time of execution these tasks. Tasks were videos with different duration time (1 s, 2 s, 5 s, 10 s). These values were chosen to get appropriate calculation time and see the difference between existing scheduling policies. What is important, the same list of tasks was executed with every policy to get results that can be compared. Characteristic of videos is provided in Table 3.

According to many experiments and tests, there was investigated that the best stability is achieved, when tasks are initialized in a row with some specific time between initialization of next task. The time between initialization of every task was set as 15 s. This is time that system need to set up connection by Master component between User and Actor, where Task Executor is initialized. With time smaller than 15 s, system was working unstable, with commonly crashed applications and errors during set up connection. That was investigated in empirical way, during huge number of test and tries.

What have to be said, the same mechanism was used with every policy (Random, Round Robin and Fuzzy) in order to achieve results that can be compared. Also, there was found out that running video analysis is relative difficult to IoT devices, and also system was working unstable with huge number of tasks executed in the same time. According to that, limit of 9 tasks was used to achieve good stability of system. This number also was investigated as result of many tests that were executed and it is Author own experience. With limit of 9 tasks on the same time, system was working appropriately. The higher this number was; the sooner system was crashed.

Sequence and time of every task was chosen in random way, but the same sequence was used with every policy to get appropriately results that can be compared.

Every video was used 6 times to get 24 different tasks. Time measurement methods that was chosen to analyze how existing and new policy are working were makespan and flowtime.

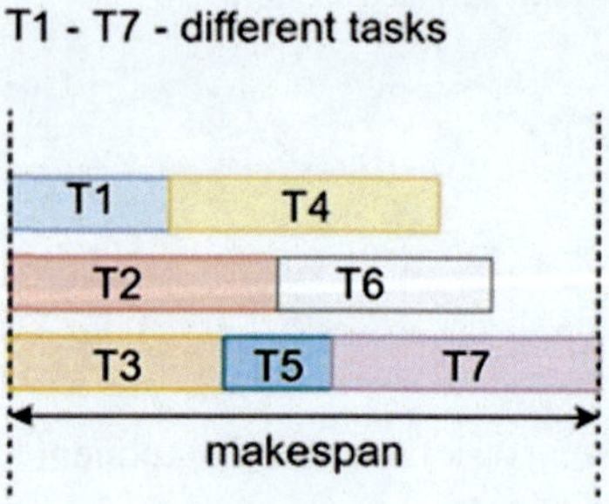

Fig. 7 Makespan

– Makespan. In this case, measured calculation time was difference between time when execution of first task was started and execution of last task was finished. This method showing time that was needed to execute all 24 tasks. This point of view can be important for service provider, because this is time when whole system is working. That means optimization of this time can be better for provider. Schematic diagram of makespan measurement method is presented in Fig. 7.
– Flowtime. Flowtime measurement method is summed up execution time of every single task. That is interesting from end user's point of view. Optimization of this time can leads to situation, when results of every single task are going to be calculated faster. That means end user can get results that he is expecting in shorter time. Schematic diagram of flowtime measurement method is presented in Fig. 8.

Selected two methods present two points of view - service provider and end user. Therefore, optimization of both times should be a goal. Nevertheless, bearing in mind that developed system is oriented on Internet of Things, user's point of view is especially important. Achieving faster results is significant in systems working in real-time or networks very sensitive to delays.

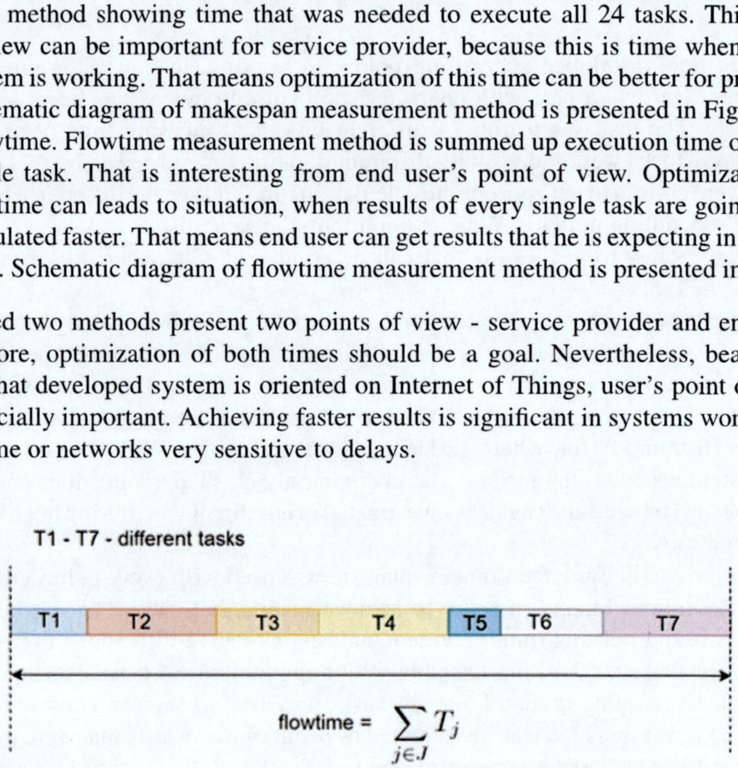

Fig. 8 Flowtime

Table 4 Obtained results

Scheduling policy	Makespan	Flowtime
Round Robin	31 m, 55 s	4 h, 19 min
Random	32 m, 05 s	4 h, 20 min
Proposed	31 m, 45 s	4 h, 13 min

5.3 Experimental Results

Experimental results are presented below. First, theoretical values of utilization (just simple test set) and performance to show, that developed fuzzy system is taking advantages of resources.

Developed scheduling policy based FRBS provides opportunity to achieve better resources management and faster time of calculation, because this policy is taking advantages about system resources and can provide tasks to devices that are less busy than another one. This knowledge and rules made whole system more efficient and faster. Comparison of executing time with described policies measured in terms of makespan and flowtime methods is presented in Table 4.

It is important to highlight that 24 tasks have been executed in sequence which was generated in a random way. The same sequence was used with every policy (Random, Round Robin and Fuzzy policy). Then, new random sequence was generated and use in the same way (once, with every policy). This process has been executed 30 times, and final results are presented in Table 4 as the average of these 30 experimental results. Simple example of random sequence with duration time described in seconds is presented below: random sequence = 1, 5, 10, 10, 2, 5, 1, 5, 10, 2, 1, 10, 2, 10, 2, 5, 5, 2, 1, 1, 2, 1, 5, 10

Implemented solution is working correctly, execution time is better, but not as much as it can be achieved. This circumstance simply shows that network was projected and deployed correctly and the FRBS is working appropriately.

6 Conclusions and Future Actions

Developed infrastructure has been made with implementation of real devices regarding a color tracking application in order to check the feasibility of AI-based docker container scheduling approaches in fog computing. Created system used many of them to create appropriate environment to tests, but also as a simulation of real systems where developed solution can be implemented. According to goals of this work, developed infrastructure was appropriate, taking into account that creation of a new scheduling policy and analysis of existing policies working with different Actors were ones of purposes. Components not directly related to the assumed goals could be implemented on the same device (PC). At the same time, the framework used in developed scenario allows the system to be decomposed into smaller components (such as Raspberry Pi), which is a great advantage of a decentralized environment based on containerization.

Developed algorithm implemented in schedule policy based on Fuzzy Rule-Based System is running correctly. It has been tested in resource-intensive complicated environment to analyze its behavior in real system with huge number of tasks. The result is that knowledge about resources and network state can be used to get better efficient system and achieve shorter time needed to finish tasks. In comparison with policies that do not use knowledge about resources and utilization of devices, developed solution have strong advantage. Additionally, results show that calculated makespan and flowtime are smaller when using the proposed approach based on FRBS.

In future, developed system can be optimized, especially fuzzy rules. To achieve this, optimization using machine learning techniques can be used. Some of them are techniques based on Swarm fuzzy systems approaches [10–13]. In addition, better knowledge about utilization of devices and network state can be used, for example implement more inputs to Fuzzy System. Some examples of new variables that can be inputs in Fuzzy System are CPU/memory utilization peak, energy consumption or bitrate of network link.

Another way to upgrade developed system is decompose PC computer (with Master, RemoteLogger and Database components) to another IoT device working as Master and RemoteLogger with Database working in Cloud. In this case, real Fog Computing system (truly in line with the paradigm of Fog Computing) will be developed. What have to be said, in this scenario strong security is relevant. This can be provided for example by using VPN technology.

References

1. Lin, J., Yu, W., Zhang, N., Yang, X., Zhang, H., Zhao, W.: A survey on internet of things: architecture, enabling technologies, security and privacy, and applications. IEEE Internet Things J. **4**(5), 1125–1142 (2017)
2. De Donno, M., Tange, K., Dragoni, N.: Foundations and evolution of modern computing paradigms: cloud, IoT, edge, and fog. IEEE Access **7**, 150936–150948 (2019)
3. OpenFog Consortium. OpenFog architecture overview. White Paper, 2016. OPFWP001, 216
4. Tuli, S., Mahmud, R., Tuli, S., Buyya, R.: Fogbus: a blockchain-based lightweight framework for edge and fog computing. J. Syst. Softw. **154**, 22–36 (2019)
5. An, J., et al.: EiF: toward an elastic IoT fog framework for AI services. IEEE Commun. Mag. **57**(5), 28–33 (2019)
6. Deng, Q., Goudarzi, M., Buyya, R.: Fogbus2: a lightweight and distributed container-based framework for integration of IoT-enabled systems with edge and cloud computing. In Proceedings of the International Workshop on Big Data in Emergent Distributed Environments, pp. 1–8 (2021)
7. Cord, O.: Genetic fuzzy systems: evolutionary tuning and learning of fuzzy knowledge bases, vol. 19. World Scientific (2001)
8. Goudarzi, M., Deng, Q., Buyya, R.: Resource management in edge and fog computing using FogBus2 framework. arXiv preprint arXiv:2108.00591 (2021)
9. Seddiki, D., et al.: Sustainable expert virtual machine migration in dynamic clouds. Comput. Electr. Eng. **102**, 108257 (2022)

10. García-Galán, S., Prado, R.P., Expósito, J.E.M.: Swarm fuzzy systems: knowledge acquisition in fuzzy systems and its applications in grid computing. IEEE Trans. Knowl. Data Eng. **26**(7), 1791–1804 (2013)
11. Prado, R.P., Garcia-Galán, S., Exposito, J.M., Yuste, A.J.: Knowledge acquisition in fuzzy-rule-based systems with particle-swarm optimization. IEEE Trans. Fuzzy Syst. **18**(6), 1083–1097 (2010)
12. García-Galán, S., Prado, R.P., Expósito, J.M.: Rules discovery in fuzzy classifier systems with PSO for scheduling in grid computational infrastructures. Appl. Soft Comput. **29**, 424–435 (2015)
13. Prado, R.P., Expósito, J.E.M., García-Galán, S.: Flexible fuzzy rule bases evolution with swarm intelligence for meta-scheduling in grid computing. Comput. Inform. **33**(4), 810–830 (2014)

Air Pollution Monitoring and Information Distribution System

Janusz Bobulski, Sabina Szymoniak, and Kamila Pasternak

Abstract The impact of air pollution on human health has been studied for decades. Unfortunately, each year brings new evidence of the harmfulness of living in a contaminated environment. The smallest particles of pollutants penetrate the entire body and affect a person already at the stage of the formation of reproductive cells. The spectrum of diseases that can develop from breathing polluted air is broad. Therefore, it is necessary to monitor the state of the air and warn of exceeding the standards, which will allow action in a specific place. In the article, we present a pollutant monitoring system that provides for their measurement and the propagation of information about their level among people in a given area. The system uses IoT technologies and the Internet.

1 Introduction

Due to rapid urbanisation and industrialisation, air pollution has become a global issue. Many countries are faced with the problem of exceeding recommended national limits. Bad air quality is Europe's most significant environmental health risk and significantly impacts the European population. Air pollution causes serious health problems such as lower respiratory infections, trachea, bronchus, lung cancers, ischaemic heart diseases and strokes. According to the WHO [1], the number of ambient air pollution-attributable deaths in 2019 exceeded 4,100,000 worldwide, 500,000 in Europe and 36,000 in Poland. Outdoor air pollution occurs in an open environment, covering the entire atmosphere. Air pollution is caused by different air pollutants emitted to the atmosphere from automobiles, chemical industries, and fossil fuel burning. The outdoor air pollutants are mainly composed of nitrogen oxides (NO_x), nitrogen dioxide (NO_2), sulfur dioxide (SO_2), ozone (O_3), carbon monoxide (CO), hydrocarbons and particulate matter (PM) of different particle sizes [2].

J. Bobulski (✉) · S. Szymoniak · K. Pasternak
Czestochowa University of Technology, Częstochowa, Poland
e-mail: januszb@icis.pcz.pl

R. Burduk et al. (eds.), *Progress on Pattern Classification, Image Processing and Communications*, Lecture Notes in Networks and Systems 766,
https://doi.org/10.1007/978-3-031-41630-9_18

It should be mentioned that external air pollution is not the only problem. It should be emphasised that internal air pollution is just as severe an issue. Exposure to indoor air pollutants can also lead to various adverse health outcomes. An example of a place where internal air pollution can occur is a household that relies on the use of polluting fuels and devices causes. The number of household air pollution deaths in 2019 exceeded 3,200,000 worldwide and 150,000 in Europe [3]. Indoor air pollution, which can be found in offices, hospitals, schools, libraries, entertainment areas, public transport vehicles, etc., is classified as the air pollution of indoor regions [4]. Major indoor air pollutants include NO_x, SO_2, O_3, CO, $carbondioxide(CO_2)$, volatile and semi-volatile organic compounds VOCs, PM, radon and microorganisms.

Because increasing air pollution is such a severe global issue, it seems reasonable to design a system for monitoring and publishing real-time air quality information. One of the most suitable and practical approaches can be the Internet of Things (IoT). It is a network where physical objects are linked to each other or to larger systems. Such a network gathers huge amounts of data from different devices and converts it into useful information. More details about IoT can be found, e.g. in [5].

Numerous IoT-based air quality monitoring systems have been proposed using micro-sensors for data collection. Studies on indoor and outdoor air pollution monitoring systems can be found in the literature. An example of the former is presented in [6]. A low-cost, portable IoT-based and real-time monitoring system for measuring ambient air quality is prosed in this research. The authors used the GSP32 module with built-in WiFi in the e-nose system for indoor air quality monitoring. The detection unit includes the sensors GP2Y1010AU, MH-Z14, MICS-4514 and DHT22, which measure air quality parameters such as CO2, CO, PM10, NO2, temperature and humidity. Another IoT-based indoor environmental quality assessment system based on an Arduino Uno module and low-cost sensors, such as DHT22, was proposed in [7]. The authors followed the Do-It-Yourself approach and used a 3D-printed nano Environmental Monitoring System. In [8], the sensors Raspberry Pi and MICS-4514 were used to create an environmental monitoring system, which enabled indoor and outdoor assessment. The system allows measuring multiple ecological parameters such as temperature, humidity, light, noise level, CO, and NO2.

An example of a system monitoring outdoor air pollution is proposed in [9]. The design and implementation of IoT sensor device is presented in this work. The authors adopted both fixed and moving IoT sensor devices. They used i.a. Arduino mega, GPS Sensor and Raspberry Pi to create the real-time localised air quality monitoring and prediction IoT sensing network. This study adopted machine learning algorithms with real-world data to develop an air quality prediction model. The paper [10] proposes a prototype of a low-cost IoT system for Air Quality Index (AQI) monitoring. The authors use i.a. Raspberry Pi module, a low-cost PM sensor and some environmental sensors to measure different parameters such as temperature, humidity and pressure. It is worth to be mentioned that authors use the same three-layered hierarchical distributed architecture for traffic flow monitoring. In [11], the open-source IoT platform collects air quality data from indoor and outdoor places. This study used GPS-based drone technology to detect and analyse air pollutants. The detection of LPG (CO2, O3, SO2), CO and NOx concentrations were realised

Fig. 1 System structure

by MQ2, MQ7 and MQ135 gas sensors. A particulate matter (PM) detector was used to monitor dust pollutants. Another work that implemented a drone to monitor air pollution is [12]. This study used a Raspberry Pi Zero microcontroller and different cost-effective sensing devices to detect concentrations of dangerous gaseous species and particulates.

2 Proposed System

2.1 System Structure

Air quality monitoring is a very important aspect of everyday life. Pollutants in the air can negatively affect human life processes and contribute to health deterioration and even death.

In this article, we propose the infrastructure of a system whose task is to monitor air quality and notify interested residents of a potential threat. The air quality monitoring system is shown in Fig. 1.

The system has three layers: data collection, end devices and communication.

The data collection layer is responsible for collecting and processing air quality data. One or more autonomous vehicles work here. Autonomous vehicles (AV) have air pollution sensors and high-performance computing servers. AVs move along specific urban routes to collect information on air quality, process it and publish it for customers in the form of statistical data and a map of a given area with marked points for measuring air pollutant concentrations.

The end-device layer brings together all devices, such as smartphones or tablets, whose users are interested in information about air quality. These devices must be equipped with a client application that allows you to log on to the server and download information about the concentration of hazardous substances in the air from the server.

Fig. 2 Vehicle used in experiment

The communication layer is responsible for the flow of communication between end devices and the data server. During communication, end devices use the Message Queue Telemetry Transport (MQTT) protocol [15]. It is a lightweight and reliable data transmission protocol designed for devices that can communicate using a limited transmission speed. MQTT is based on the publish and subscribe model. This means that the devices in the system have specific roles defined. The autonomous vehicle is a publisher that posts information on pollution in a given agglomeration (maps, figures). Each end device acts as a Subscriber. Subscriber reads (subscribes to) a report on a specific topic published by the Publisher. The case can be mapped with marked information about air quality and a list of all concentrations of all measured substances or information about one of them. The server located in the autonomous vehicle acts as the MQTT-Broker. Clients, i.e. terminal devices, connect to the Broker to download information of interest to them. In contrast, autonomous vehicles [13] connect to a broker to publish data.

2.2 Vehicle

To ensure the correctness of measurements, interference sources should be eliminated during experiments. In the discussed case, interference in measuring air pollution could be caused by the measuring vehicle itself [14]. That's why we used a Nissan Leaf electric vehicle in our experiment (see Fig. 2).

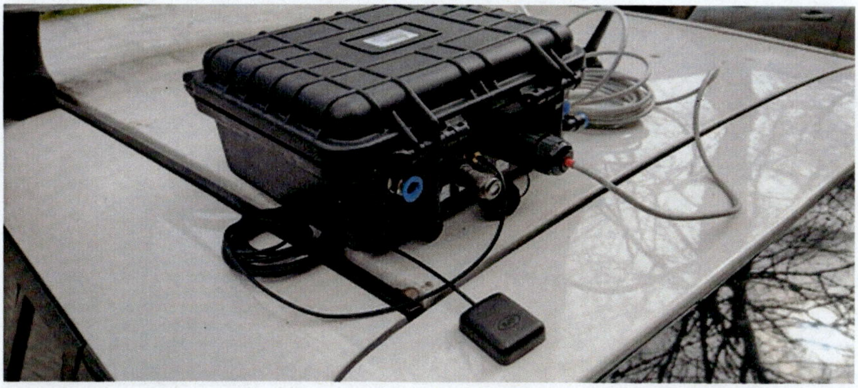

Fig. 3 Instrument used in experiment

2.3 *Instruments*

The measuring device consists of two modules built into one Peli-type case:

- a measuring module with a built-in suspended dust measurement system,
- a measuring module with gas sensors working in a pump system.

Technical specification of the measuring instrument:

- PM10 dust range from 0 to 1999.9 $\mu g/m^3$, sensitivity 0.3 $\mu g/m^3$
- PM10 dust range from 0 to 999.9 $\mu g/m^3$
- VOC Volatile Organic Compounds PID 10.6eV detection range: 0 ppm to 50 ppm, photoionization method,
- CO range from 1 to 500 ppm, the electrochemical method
- NO2 range from 0 to 20 ppm, the electrochemical method
- NO range 0 to 20 ppm, the electrochemical method
- SO2 range from 0 to 50 ppm, the electrochemical method
- temperature $-40\,°C$ to $85\,°C$, resolution $0.1\,°C$
- humidity from 0 to 100
- pressure from 300 hPa to 1100 hPa, accuracy 100 Pa
- GPS position (GNSS) accuracy up to 1.5 m
- GSM/3G/LTE module
- WiFi 2.4/5.8 GHz 802.11an

The measuring instruments were integrated into a suitcase and then placed with magnets on the research vehicle. The power supply for the measuring set (grey cable in Fig. 3) came from the cigarette lighter socket. The set had a GPS module (black cable in Fig. 4).

Fig. 4 Instrument on Vehicle used in experiment

Table 1 The data of air quality.

Time	LAT	LON	ALT	PM10	PM25	NO2	SO2	NO	PID	Temp	Press	Hum
14.11.2022 08:46	51.253555	22.563112	0	9.2	5.4	0.129	0	0.097	0.199	8.88	1007.26	68.89
14.11.2022 08:45	51.253555	22.563112	0	9.2	5.4	0.129	0	0.097	0.199	8.88	1007.26	68.89
14.11.2022 08:44	51.253555	22.563112	179.6	8.1	4.3	0.12	0	0.11	0.195	8.76	1007.06	68.9
14.11.2022 08:43	51.253135	22.560657	183.2	6.9	3.7	0.136	0	0.223	0.198	9.04	1007.17	68.27
14.11.2022 08:41	51.251875	22.551793	191.4	9.9	4.8	0.078	0	0.123	0.187	9.51	1006.38	66.89
14.11.2022 08:40	51.250048	22.552458	198.5	7.6	4.4	0.142	0	0.171	0.182	9.63	1004.89	65.52
14.11.2022 08:39	51.249043	22.554722	192.2	11.2	4.9	0.249	0	0.858	0.207	9.68	1005.28	65.99
14.11.2022 08:38	51.249638	22.55818	199.4	8.6	4.6	0.116	0	0.284	0.183	9.82	1004.94	65.06
14.11.2022 08:37	51.251273	22.558928	182.1	9	4.8	0.097	0	0.353	0.181	9.94	1006.06	64.2
14.11.2022 08:36	51.253255	22.561388	172.2	9.1	4.7	0.071	0	0.106	0.173	10.21	1007.28	62.32
14.11.2022 08:35	51.253255	22.561388	172.2	19.5	10.9	0.076	0	0.259	0.176	10.61	1007.33	61.47
14.11.2022 08:34	51.253093	22.563277	171.9	7.9	4.7	0.157	0	0.12	0.172	11.1	1007.52	58.85
14.11.2022 08:33	51.253477	22.56299	177.8	7.1	3.8	0.073	0	0.101	0.17	11.86	1007.43	55.57

2.4 Results

The presented pollution measurement system collects current data on air quality in the form presented in Table 1. They can then be made available to subscribers of the service offered by the system. The data can be presented in a graphical form (Fig. 5) where the circle's size shows the pollution level, and the red colour exceeds the standards. Another method of presenting the results is displaying numerical values (Fig. 6) at a given location.

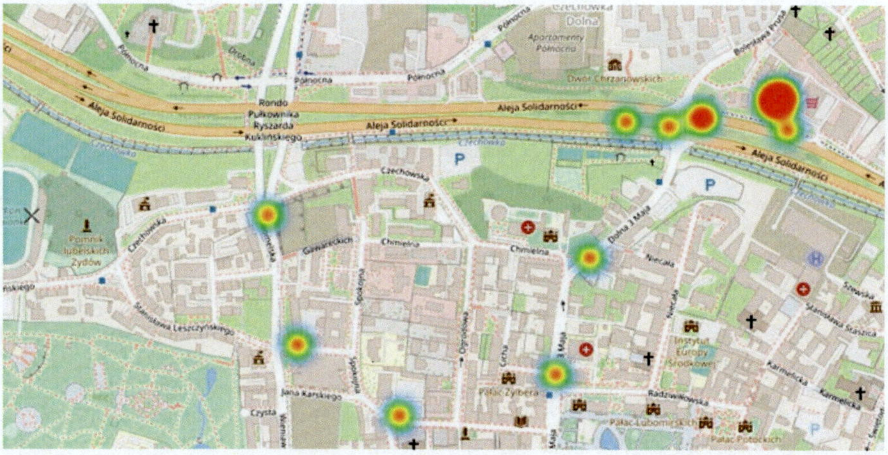

Fig. 5 Results of measurement in form of maps and colors

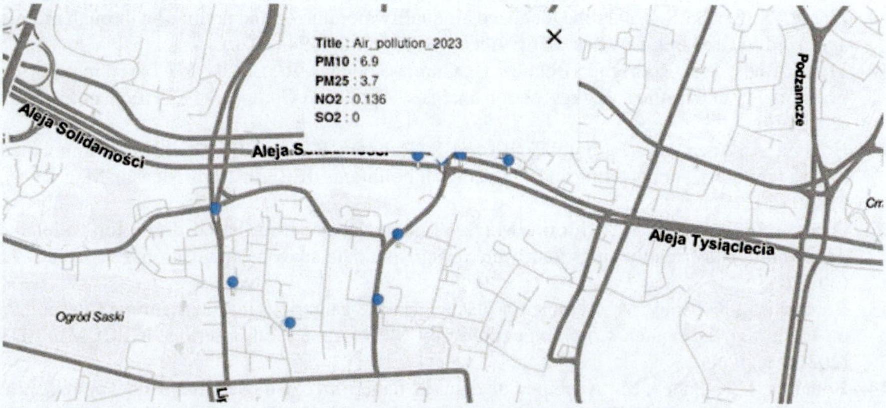

Fig. 6 Results of measurement in form of maps and numbers

3 Conclusion

Benefits of air quality monitoring. Air quality monitoring is an essential task for health and the environment. Today, more and more countries, municipalities and communities attach great importance to air quality monitoring to better understand its composition and possible harmful influences. The article shows the air quality monitoring system with the results' distribution. Controlling the state of the air allows you to detect sources of pollution and eliminate them. These activities reduce local air pollution and improve residents' quality of life. In addition, the same actions minimise greenhouse gas emissions and save money in many cases.

References

1. https://www.who.int/data/gho/data/indicators/indicator-details/GHO/ambient-air-pollution-attributable-deaths . Accessed 23 Mar 2023
2. Kampa, M., Castanas, E.: Human health effects of air pollution. Environ. Pollut. **151**(2), 362–367 (2008)
3. https://www.who.int/data/gho/data/indicators/indicator-details/GHO/household-air-pollution-attributable-deaths . Accessed 23 Mar 2023
4. De Gennaro, G., et al.: Indoor air quality in schools. Environ. Chem. Lett. **12**, 467–482 (2014)
5. Gubbi, J., Buyya, R., Marusic, S., Palaniswami, M.: Internet of Things (IoT): a vision, architectural elements, and future directions. Future Gener. Comput. Syst. **29**, 1645–1660 (2013)
6. Tstan, M., Gokozan, H.: Real-time monitoring of indoor air quality with internet of things-based E-nose. Appl. Sci. **9**(16), 34–35 (2019)
7. Salamone, F., Danza, L., Meroni, I., Pollastro, M.: A low-cost environmental monitoring system: how to prevent systematic errors in the design phase through the combined use of additive manufacturing and thermographic techniques. Sensors **17**(828) (2017)
8. Alvarez-Campana, M., López, G., Vázquez, E., Villagrá, V., Berrocal, J.: Smart CEI Moncloa: an IoT-based platform for people flow and environmental monitoring on a Smart University Campus. Sensors **17**, 28–56 (2017)
9. Zhang, D., Woo, S.S.: Real time localized air quality monitoring and prediction through mobile and fixed IoT sensing network. IEEE Access **8**, 89584–89594 (2020)
10. Martín-Baos, J.Á., Rodriguez-Benitez, L., García-Ródenas, R., Liu, J.: IoT based monitoring of air quality and traffic using regression analysis. Appl. Soft Comput. **115**, Article no. 108282 (2022)
11. Hossain, A., Anee, M.J., Faruqui, R., Bushra, S., Rahman, P., Khan, R.: A GPS based unmanned drone technology for detecting and analyzing air pollutants. IEEE Instrum. Measur. Mag. **25**(9), 53–60 (2022)
12. Fazio, R.D., Dinoi, L.M., Vittorio, M.D., Visconti, P.: A sensor-based drone for pollutants detection in eco-friendly cities: hardware design and data analysis application. Electronics **11** (2022)
13. Bobulski, J., Kubanek, M.: Vehicle for plastic garbage gathering. In: International Conference on Electrical, Computer, Communications and Mechatronics Engineering, ICECCME 2021 (2021)
14. Bobulski, J., Kubanek, M.: A method of cleaning data from IoT devices in big data systems. In: Proceedings of the 2022 IEEE International Conference on Big Data, Big Data, pp. 6596–6598 (2022)
15. Zeghida, H., Boulaiche, M., Chikh, R.: Securing MQTT protocol for IoT environment using IDS based on ensemble learning. Int. J. Inf. Secur. **22**, 1–12 (2023)

Detection of Conspiracy Theories from Tweets Using NLI-Based Zero Shot Text Classification Models

Weronika Jakubowska, Magdalena Kociołek, and Dominik Kurowski

Abstract In recent years, conspiracy theories have gained popularity, and during the COVID-19 pandemic, they had a significant impact on people's lives. Some individuals claimed that COVID-19 was a fabrication or bio-weapon created by specific people or governments. It can be challenging for social media users to differentiate between genuine news and conspiracy theories. To address this issue, a study was conducted to determine if tweets with conspiracy theories could be distinguished from neutral tweets. Twitter was searched, and deep learning methods were applied to examine the tweet content, sentiment, and user reactions. The study found that real-time monitoring of these factors could be an effective way to prevent the spread of misinformation on social media.

Keywords Twitter · Conspiracy theory · Zero-shot learning · Machine learning · Deep learning · Social media

1 Introduction

Conspiracy theories cover a wide range of topics, from aviation and accidents to economics and deaths. According to the Cambridge Dictionary[1], a conspiracy theory is a belief that powerful individuals have secretly planned an event or situation. Another definition [2]states that conspiracy theories reject the standard explanation for an event and instead attribute it to a covert group or organization carrying out a secret plot. These theories are widespread on both television and the Internet, with social media serving as a platform for their rapid spread. Conspiracy theories often relate to popular or inexplicable events, such as the terrorist attack on the World Trade Center, which gave rise to numerous theories about what "really" happened. More recently, a plethora of conspiracy theories have emerged regarding the coronavirus

W. Jakubowska (✉) · M. Kociołek · D. Kurowski
Wroclaw University of Science and Technology, 27 Wybrzeże Wyspiańskiego st., 50-370 Wrocław, Poland
e-mail: jakubowska_weronika22@wp.pl

R. Burduk et al. (eds.), *Progress on Pattern Classification, Image Processing and Communications*, Lecture Notes in Networks and Systems 766,
https://doi.org/10.1007/978-3-031-41630-9_19

pandemic and vaccination. Despite often contradicting one another, these theories have the potential to cause significant harm. Therefore, it is crucial to develop automated models that can detect conspiracy theories on social media platforms. Doing so will prevent nefarious actors from manipulating vulnerable communities, thereby reducing the potential for negative social and political outcomes. To address this issue, we conducted a study analyzing data from Twitter to determine whether conspiracy theories are identifiable by their content and people's reactions. We developed a Zero-Shot NLI classification model to distinguish tweets as conspiracy theories or not and examined how people respond to these theories. Additionally, we created a model to detect controversial tweets and investigated whether conspiracy theories are more controversial than other news. This study and model results demonstrate that it is possible to differentiate between conspiracy theories and factual tweets on Twitter based on their content and users' reactions.

2 Related Works

There have been many studies on identifying conspiracy theories, covering a wide range of topics. Some focus on classifying text, while others prioritize creating high-quality training sets. In 2022, a dataset of 3,100 manually annotated tweets was created at Carnegie Mellon University[6], containing information on various domains such as climate change, criminal trials, and public health. Each tweet was labeled based on whether it contains a conspiracy theory, the topic, and stance on the theory. Different models were used to detect conspiracy theories, such as Majority classifier, BERT, ALBERTA, RoBERTa, and DistilBERT, with RoBERTa yielding the best results. Other methods used include random forest classification models[7] and support vector machine methods[8]. Word embeddings were also utilized to distinguish between non-conspiracy theory content and conspiracy-related content in scientific publications[9].

3 Proposed Data Pipeline

In this section, we describe our dataset and used models.

3.1 Data Collection

In our study, we selected specific conspiracy theories from a list on Wikipedia[3] and developed thematic keywords to better understand their content. Selected conspiracy theories and themes:

Table 1 Number of tweets in collected dataset

Keyword	tweets
plandemia	18 128
black helicopters	55 839
predictive programming	85 950
malaysia airlines flight MH370	88 401
reptilians	100 049
scamdemic	163 920
new world order	229 019
freemasonary	245 655
conspiracy theories	405 159
brainwashing	500 002
illuminati	586 991
wtc	608 096
vaccination	625 483
jfk	664 184
chemtrails	669 658
conspiracy	890 001

1. Black Helicopters,
2. Chemtrails,
3. Freemasonary,
4. Illuminati,
5. John F. Kennedy's death,
6. Malaysia airlines flight MH370.
7. pandemy and vaccines,
8. Reptilians,
9. World Trade Center attack 9/11/01.

We used the snscrape Python library Twitter module[4] to collect publicly available data from over 5 million unique Twitter tweets containing specific keywords related to conspiracy theories. Each tweet was labeled with its corresponding keyword, allowing for separate analysis of each theory and comparison between them. Table 1 shows the number of tweets in the dataset for each keyword.

In addition to the tweet text content, we were able to scrape the publication date, username, and reactions for all tweets, including likes, replies, and retweets. To ensure the accuracy of the data, we removed any images, URLs, emoticons and symbols, usernames, hashtags, and non-English text strings from the tweets.

3.2 Models

Controversial Detection

In recent years, large language models have shown impressive performance on a range of natural language processing tasks. One promising application of these models is their use as zero-shot classifiers for detecting controversial content. Zero-shot models are not fine-tuned for a specific task but rather rely on general knowledge learned from pre-training on large datasets. To achieve our objective, we used a mDeBERTa-v3-base-mnli-xnli model, which has been pre-trained to solve the NLI problem, i.e., recognizing the relationship between two given texts. The primary function of an NLI model is to compare two texts and determine whether there is an entailment (implies), neutral (no relationship), or contradiction (contradicts). By utilizing this feature, we can transform any classification problem into an NLI task. This can be accomplished by comparing the tweet with a statement that asserts its controversial nature and examining the NLI model's output, for example: [Tweet content] | ["This tweet is controversial"]. If the model outputs "entailment", it means that the tweet is controversial, whereas "neutral" or "contradiction" suggests the opposite. Therefore, we can leverage the power of pre-trained NLI models to zero-shot classify tweets based on their controversial nature. This technique enables us to utilize the vast knowledge of the model, even for tasks it was not specifically trained on. To evaluate our approach, we perform experiments on a dataset of controversial texts. Our results show that our method achieves sufficient performance compared to supervised models, without the need for fine-tuning or additional training.

Conspiracy Theory Detection

In addition to examining the controversial nature of tweets, we tackled the task of detecting whether a tweet is a conspiracy theory. In our work we trained a large language model RoBERTa, on the "Hoaxes and Hidden agendas: A Twitter Conspiracy Theory" dataset. The dataset contains 3100 tweets, manually labeled by experts as either conspiracy theories or not. The RoBERTa model was fine-tuned on this dataset using a binary classification task, with the output layer being a Sigmoidal layer. The output score of the Sigmoidal layer was used as a measure of the probability that a given tweet was a conspiracy theory. To determine a threshold for classifying a tweet as a conspiracy theory, we analyzed the score distribution of the training data and determined that a threshold of 0.75 provided the best balance between precision and recall.

Results

Our benchmark is a dataset created at Carnegie Mellon University [6]. Our RoBERTa model achieved an F1 score of 0.74 on the test set, indicating its relatively good accuracy in detecting conspiracy theories (Table 2).

The comparison with results described in the benchmark publication present Table 3. We obtained results better than Majority Classifier and similar to BERT and ALBERT. It is worth noting that our model was trained on different data than the models from the publication, and our tweets covered more conspiracy theories and

Table 2 The results of our model to detect conspiracy theories

Label	Precision	Recall	F1-Score
Not conspiracy theory	0.88	0.81	0.85
Conspiracy theory	0.57	0.69	0.63

Table 3 Compare our model with others models [6]

Model	Precision	Recall	Macro-F1
Majority classifier	0.38	0.50	0.43
BERT	0.81	0.80	0.80
ALBERT	0.73	0.78	0.75
RoBERTa	**0.86**	0.80	**0.81**
DistilBERT	0.80	**0.81**	0.80
Our model	0.73	0.75	0.74

different periods. The biggest advantage of our solution is that we achieved similar results with minimal data for tuning, while the disadvantage of the solutions from the benchmark publication is that they required complete model training, which took up significant time.

4 Analysis

4.1 Data Analysis

When analyzing tweets related to conspiracy theories, we grouped them by their labels and analyzed them as groups. We then computed the daily number of tweets and active users for each keyword. This allowed us to treat conspiracy theories like time-series data. Our study revealed that conspiracy theories can be categorized into two types: seasonal theories, such as the theory about the World Trade Center collapse, and theories with constant user involvement, such as Predictive Programming. Examples are shown in Fig. 1 and Fig. 2.

According to our analysis, there are over 7,000 new conspiracy posts and 3,000 new active users every day. In terms of user reactions to conspiracy theories, we found that New World Order had the most likes and retweets, while Chemtrails had the least likes, retweets, and replies. Reptilians had the most replies to a tweet. Interestingly, among non-controversial tweets, vaccinations had the highest number of replies to a tweet. It's clear that each conspiracy theory elicits a unique response from Twitter users.

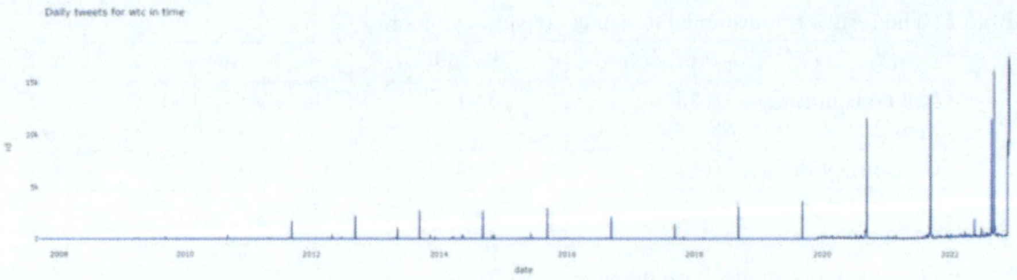

Fig. 1 Daily number of tweets for World Trade Center conspiracy theory

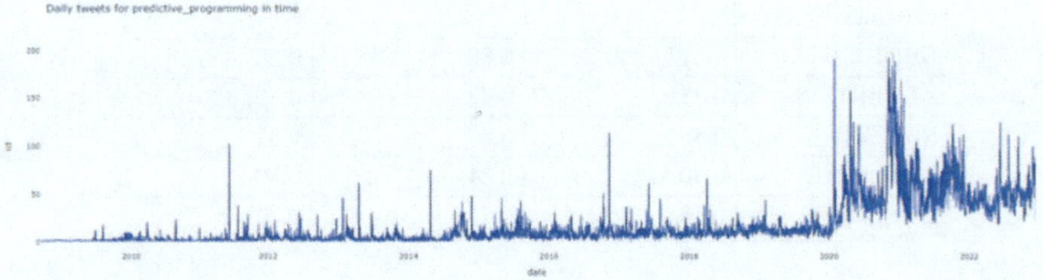

Fig. 2 Daily number of tweets for predictive programming conspiracy theory

4.2 Controversy Analysis

Research has been conducted on the controversial nature of tweets.

Figure 3 shows the percentage of controversial tweets for each conspiracy theory analyzed. The results indicate that conspiracy theories are not necessarily more controversial than other types of tweets.

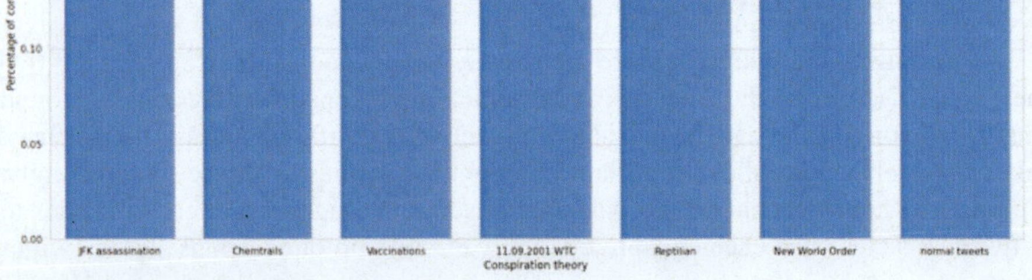

Fig. 3 Percentage of controversial tweets for each theory

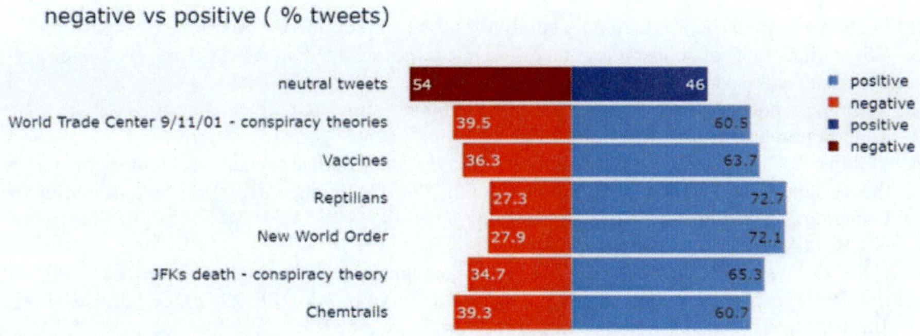

Fig. 4 Sentiment of tweets

4.3 Sentiment Analysis

To analyze the sentiment of tweets, we used VaderSentiment SentimentIntensityAnalyzer [5]. Based on the aggregated analysis, it appears that there is a difference in sentiment between neutral tweets and conspiracy tweets. Figure 4 displays the results of this analysis. In conclusion, it seems that neutral tweets have a tendency towards negativity, while conspiracy tweets are dominated by positive content.

5 Conclusions

Conspiracy theories can have a divisive impact on societies, often spreading disinformation or conflicting with societal values. Unfortunately, there are no clear warning signs on social media platforms, making it difficult for users to distinguish between conspiracy and fact. This research has examined whether machine learning methods could help in this regard. The Zero-Shot NLI model was used in this study, with a precision rate of 0.72, which indicates whether a tweet is a conspiracy theory or not. The analysis of user reactions revealed that conspiracy theories are no more controversial than neutral tweets, but they tend to contain more positive messages. Furthermore, these findings provide additional insight into each conspiracy theory. Overall, our study and model results demonstrate that it is possible to differentiate between conspiracy theories and factual tweets on Twitter based on their content and users' reactions.

References

1. Cambridge Dictionary. https://dictionary.cambridge.org/dictionary/english/conspiracy-theory. Accessed 22 Apr 2023

2. Dictionary. https://www.dictionary.com/browse/conspiracy-theory. Accessed 22 Apr 2023
3. Wikipedia, List of conspiracy theories. https://en.wikipedia.org/wiki/List_of_conspiracy_theories. Accessed 22 Apr 2023
4. Snscrape repository. https://github.com/JustAnotherArchivist/snscrape. Accessed 22 Apr 2023
5. VaderSentiment. https://pypi.org/project/vaderSentiment/. Accessed 22 Apr 2022
6. Phillips, S.C., Ng, L.H.X., Carley, K.M.: Hoaxes and Hidden agendas: a Twitter conspiracy theory dataset. In: Companion Proceedings of the Web Conference 2022, France. Association for Computing Machinery (2022). https://doi.org/10.1145/3487553.3524665. https://dl.acm.org/doi/10.1145/3487553.3524665
7. Gerts, D., et al.: "Thought I'd share first" and other conspiracy theory tweets from the COVID-19 infodemic: exploratory study. JMIR Public Health Surveill. **7**(4), e26527 (2021). https://doi.org/10.2196/26527. https://publichealth.jmir.org/2021/4/e26527/
8. Pogorelov, K., Schroeder, D.T., Filkuková, P., Brenner, S., Langguth, J.: WICO text: a labeled dataset of conspiracy theory and 5G-corona misinformation tweets. Association for Computing Machinery, New York (2021). https://doi.org/10.1145/3472720.3483617
9. Batzdorfer, V., Steinmetz, H., Biella, M., et al.: Conspiracy theories on Twitter: emerging motifs and temporal dynamics during the COVID-19 pandemic. Int. J. Data Sci. Anal. **13**, 315–333 (2022). https://doi.org/10.1007/s41060-021-00298-6

Battery Management System for Polish Hybrid Residential Photovoltaic Power Plants

Paweł Parczyk and **Robert Burduk**

Abstract Growing share of Renewable Energy Sources (RES) in Polish electricity system has motivated the Polish government to make changes in the law. Those changes decrease the economic profit of using photovoltaic power plants, but incentivize the use of hybrid systems equipped with batteries. In this paper, the state-of-the-art battery management system, which maximizes self-consumption has been compared to our own time-based proposal, which is focused on economic profit. We used real-life measurements from a 6kWp residential power plant to compare both battery management systems. The results show that controlling batteries with our model allows us to save more money.

Keywords RES · Time series optimisation · Battery management systems · Self-consumption

1 Introduction

The continuous growth of the share of renewable energy sources (RES) especially photovoltaic (PV) in the electricity mix changed the point of view on future development of RES. More and more energy provided by unstable RES can damage the electrical system. It creates grid overvoltage and can cause electricity drops. That shown us new challenges. In order to meet the presented problems, the Polish government has altered the laws regarding residential power plants [4]. They replaced `net-metering` billing system with `net-billing`. In the new system, more market-like features are introduced in the calculation of the value of exported and

P. Parczyk (✉) · R. Burduk
Department of Systems and Computer Networks, Wrocław University of Science and Technology, Wybrzeże Wyspiańskiego 27, Wrocław, Poland
e-mail: pawel.parczyk@pwr.edu.pl

R. Burduk
e-mail: robert.burduk@pwr.edu.pl

R. Burduk et al. (eds.), *Progress on Pattern Classification, Image Processing and Communications*, Lecture Notes in Networks and Systems 766,
https://doi.org/10.1007/978-3-031-41630-9_20

imported energy by residential photovoltaic systems. The value of exported energy is based on the market price of energy.

This paper presents an analysis of a modern and widely used battery management system, which is Self-Consumption (`StBMS`) [2]. We found our problem is embedded especially in a domain of time series analysis. The problem covers forecasting of various variables changing over time and selection of the proper course in time of the variable under control.

In the Sect. 2, we present some weaknesses of `StBMS`, which occur especially in the new Polish billing system. Moreover in the Sect. 3, we propose a time-based battery management system (`TbBMS`), which surpasses `StBMS` in some points and is better suited for `net-billing`. In Sect. 4, we compare the economic impact of those two `BMSs`, on a dataset spanning data from over a half-a-year of running 6kWp `PV` power plant located near Wrocław.

2 Self-consumption Battery Management System

`StBMS` is a simple and widely used battery management system in residential `PV` power plants. It uses a battery and minimizes the exchange of the energy between grid and the household energy system. The main advantage of `StBMS` is its independence from external information apart from the current system state. It means that policy is calculated based on the current measures from the system (Algorithm 1 defines how `StBMS` works).

Algorithm 1: `StBMS` algorithm

$PV_n \leftarrow$ PV power;
$C_n \leftarrow$ Consumption (demand) power;
$Bat_n \leftarrow$ Computed battery power;
$SoC_n \leftarrow$ Computed battery State of Charge;
$n \leftarrow$ Number of samples in the timeseries;
$\Delta T \leftarrow$ Time difference between samples in timeseries;
Procedure *StBMS(PV, C)*
    $SoC_0 \leftarrow 0$;
    $Bat_0 \leftarrow 0$;
    foreach $i \in 0, \ldots, n-1$ **do**
        $PD = PV_i - C_i$;
        if $(PD > 0 \wedge SoC_i < 100\%) \vee (PD < 0 \wedge SoC_i > 0\%)$ **then**
            $Bat_{i+1} \leftarrow PD$;
        else
            $Bat_{i+1} \leftarrow 0$;
        end
        $SoC_{i+1} \leftarrow SoC_i + Bat_i \cdot \Delta T$;
    end
    return Bat, SoC

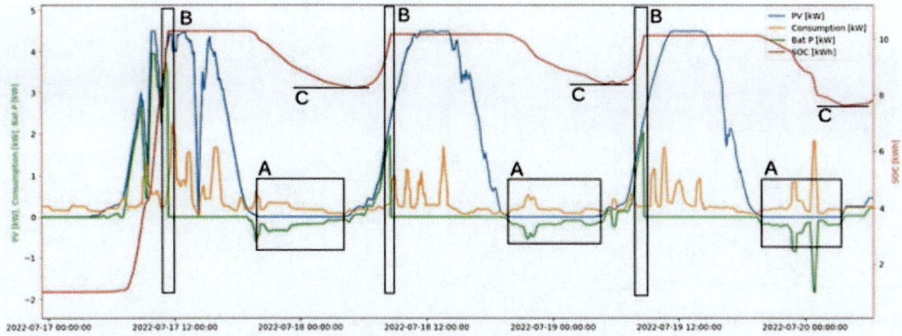

Fig. 1 A simulation of `StBMS` on a 3-day window of the middle of summer. The red line portrays the state of charge (`SoC`) of the battery, the blue line represents `PV` power, the orange line represents power consumption of the house and the green line represents the power of a battery charging or discharging (Bat).

The algorithm can be used to calculate battery power (Bat) and battery State of Charge (`SoC`). In our simulation, we made two assumptions. Battery capacity is 10 kWh and maximum charge and discharge power is 6kW. Using this algorithm we conduct a simulation. The results from three days of this simulation are shown in the Fig. 1.

The charts portray three key features. From sunrise, the `PV` power is rising, when `PV` power exceeds consumption power the battery charging starts. `StBMS` ensures that the sum of battery power and consumption is equal to or less than `PV` power. The system avoids charging from the grid. Once the battery is full, `SoC` reaches 100 % (see **B** section in the figure), and battery power decreases to 0kW. For the rest of the day, excess energy is exported to the grid. After sunset, the battery starts discharging, providing power equal to consumption. This process is visible in **A** section in the figure. At the end of the night, the battery still holds a partial charge, which is much more than required. The lowest level is shown by the line **C** at the level of about 8 kWh.

We can observe two features. First is that the battery is always charged at the beginning of the day. And at the end of the night, the battery still has some energy stored. Depending on the battery capacity and season, the remaining `SoC` can vary. This behavior decreases usefulness of the battery.

In `net-billing` exported energy is exchanged for money, and the exchange rate varies over the course of the day. The price of electricity is lowest at about 12 am and highest at about 8pm. The prices are shown in the Fig. 2.

We can check how export and import fit into the prices of energy. We can observe that most of the export is made when the energy is cheapest (see **B** section in the Fig. 3). Additionally, at the beginning of the day, no energy is exported (see **A** section in the Fig. 3). It is clearly visible that it would be better to export energy at the beginning of the day and to limit export on midday, as even if cumulative exported energy would be the same, the cumulative profit would be higher.

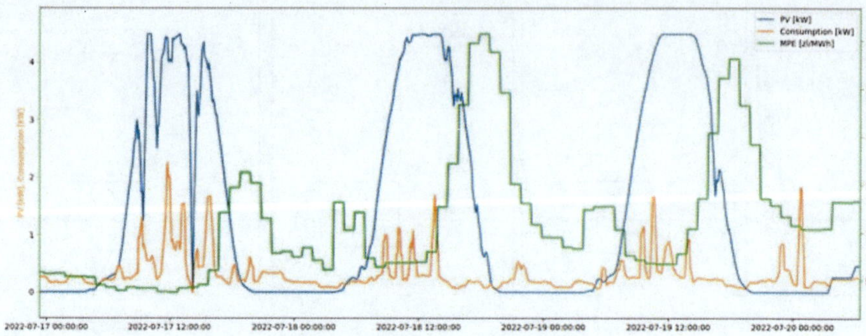

Fig. 2 A price line compare with PV power and consumption

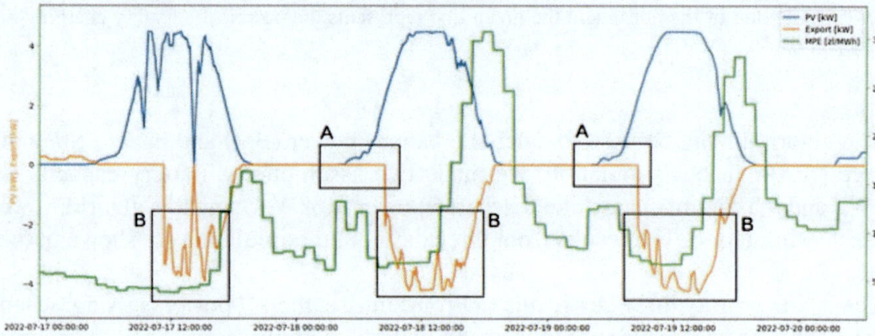

Fig. 3 A price line MPE compared with PV power and house energy consumption

Taking into account described weaknesses of StBMS we have prepared a Time-Based Battery Management System.

3 Time-Based Battery Management System

Our model is answering the weaknesses of the state-of-the-art StBMS. The goal of that model is to reduce the SoC at the end of the day, to the lowest possible value and to delay battery charging to ensure that the battery will be charged with the cheapest energy and discharged the most expensive energy. However, it is also important to ensure as much Self-Consumption as possible. This causes us some difficulties. The model should meet the following requirements:

- Firstly, the model has to identify the time when energy is the cheapest and the most expensive. To ensure that we have made some observations. Strong seasonality on MPE (see Fig. 2) is visible. There are two peaks in electricity prices - first, at 6 am, and second, in the evening. At about midday, we can observe a dip on the chart. The length of the compared period is one-day.

– Secondly, the model has to discharge the battery with some amount of energy, but it cannot discharge completely. We still expect to meet overnight demand with own energy (not from the grid). To ensure that, the model has to calculate the overnight demand. A short analysis of consumption brings similar conclusions as in the case of MPE. Consumption is characterized by strong one-day seasonality.
– Thirdly, the model has to charge the battery with the cheapest possible energy, but at the end of the day when `PV` production disappears, the battery should be fully charged regardless of changing weather.

All of the above requirements require knowledge of the values of upcoming MPE, consumption, and `PV`. To simplify the model we used an `Identity` forecasting model. It simply forecasts the same value as occurred at the same position in the last period (last day).

Regarding the formal description of the Algorithm 2, we can observe that `TbBMS` differs from `StBMS`. The default behavior is similar, but in two cases the algorithm is changed. The first case is just after the maximum MPE has been met. The algorithm forecasts the demand, and, if there is more power stored than the forecasted demand, the algorithm tries to sell excess energy. This procedure happens before the next `PV` production starts. The second case is when `PV` started, but the lowest price has not been reached. Algorithm forecasts `PV` production and if it is higher than needed to fully charge the battery it stops charging. However, it works only if the *enable_delayed_charging* flag has been set to true, otherwise, this functionality is disabled.

In the Fig. 4 a 3-day working period of `TbBMS` is visualized. It can be observed that our `BMS` meets the stated requirements. At the beginning of `PV` production, we can observe the charging is delayed (see section B in the Fig. 4). In comparison to the `StBMS` our model does not charge the battery as soon as it is possible. It rather waits until the energy is cheapest. This approach, however, has some imperfections. Sometimes, the battery won't be fully charged at the end of the day. This happens when there are significant differences between MPE in successive days. The second behavior of `TbBMS` is well visible in section **A** in the Fig. 4. The model starts emptying the battery just after the maximum price and ensures that the battery is empty for the next day's generation (see the crossing of the lines **C** and **D**).

4 Experiments

For the purpose of model evaluation we prepared an experimental protocol, according to which we conducted our experiments. First we prepared a time series containing `PV`, power demand and energy price. After that, using rolling-window scheme we fit models with window (a period preceding the considered time-point) and made calculations for the following period. Then rolling-window was moved forward and the procedure was repeated. Effectiveness of the models was evaluated using economic impact.

Algorithm 2: TbBMS algorithm

```
PV_n ← PV power;
C_n ← Consumption (demand) power;
Bat_n ← Battery power;
SoC_n ← The battery State of Charge;
fPV_k ← forecasted PV for next period of k;
fC_k ← forecasted consumption for next period of k;
PV_begin ← forecasted time of next occurrence of the beginning of the PV;
PV_end ← forecasted time of next occurrence of the end of the PV;
MPE_min ← forecasted time of the occurrence of the cheapest energy;
MPE_max ← forecasted time of the occurrence of the most expensive energy;
n ← Number of samples in the timeseries;
k ← Period of forecast. In this case one day ahead 288 samples;
ΔT ← Time difference between battery power assignment;
Procedure TbBMS(PV, C, MPE, enable_delayed_charging, battery_capacity)
    SoC_0 ← 0;
    Bat_0 ← 0;
    /* Iterate over days */
    for j ∈ 0, k, 2·k, ···, (n−1)/k do
        /* Compute forecasts for the coming day */
        fPV = Identity(PV, k, j);                       /* Compute PV forecast */
        /* Compute begin and end of production */
        PV_min ← argmin(fPV = 0) + j;
        PV_max ← argmax(fPV ≠ 0) + j;
        fMPE = Identity(MPE, k, j);                     /* Compute MPE forecast */
        /* Compute time of the cheapest and the most expensive
           energy */
        MPE_min ← argmin(fMPE) + j;
        MPE_max ← argmax(fMPE) + j;
        /* Iterate over each sample within a day */
        for (i ∈ j, j+1, j+2, ···, j+k−1) do
            Bat_{i+1} = PV_i − C_i
            /* In the morning before PV starts */
            if (MPE_max < i ∨ i < PV_begin ∧ SoC_i > Σ_i^k fC_{i−j:PV_begin−j}) then
                Bat_{i+1} ← SoC_i / (PV_begin − i) ;           /* Discharge slowly */
            /* In case of the early morning after PV starts but
               before the cheapest price */
            else if (PV_begin < i < MPE_min ∧ enable_delayed_charging) then
                if (Σ_i^k fPV_{i−j:k} > battery_capacity − SoC_i) then
                    Bat_{i+1} ← 0;                             /* Delayed charging */
                end
            end
            /* Just after the maximum price */
            else if (MPE_max < i ∧ SoC_i > Σ_i^k fC_{i−j:k} + 20% * battery_capacity) then
                Bat_{i+1} ← SoC_i / (k+j−i);                   /* Discharge slowly */
            SoC_{i+1} ← SoC_i + Bat_i · ΔT;
        end
    end
    return Bat, SoC
```

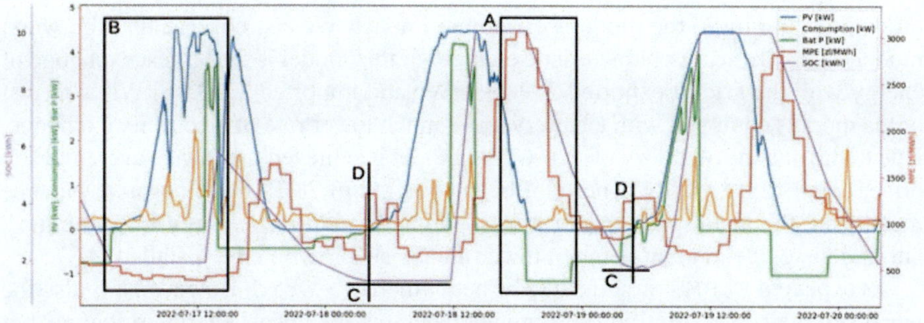

Fig. 4 A simulation of TbBMS on a 3-day window in the middle of summer. Red line portrays MPE, the blue line represents PV power, the orange is house energy consumption and green is a battery charge or discharge power (Bat), purple is a state of charge (SoC).

Table 1 Results of selected models On-Grid, StBMS, TbBMS and SBMS on half-a-year dataset taken from 6kWp residential PV power plant located near Wrocław

BMS	E. Export kWh	E. Import kWh	Income PLN	Outcome PLN	Balance PLN	Avg. Exp. Price PLN/kWh	Avg. Imp. Price PLN/kWh
On-grid	**2198.99**	2763.84	**1866.52**	3040.23	−1173.70	0.84	1.1
StBMS	1617.33	**2182.11**	1428.40	**2400.32**	−971.92	0.88	1.1
TbBMS	1776.57	2341.40	1710.03	2575.54	−865.50	0.96	1.1
SBMS	1729.68	2294.47	1716.64	2523.91	**−807.26**	**0.99**	1.1

During the experiments, we selected 4 models to be presented in the results. All of them represent alternative strategies for battery management. We used:

- On-Grid - the default installation without any battery systems.
- StBMS - State-Of-The-Art algorithm providing maximum Self-Consumption.
- TbBMS - Our own model, which has been described in this paper.
- SBMS (Sell BMS) - Variation of our model, with disabled delayed charging functionality (*enable_delayed_charging* set to false).

We used our own dataset of over half a year of observations taken from a 6kWp PV power plant. We downloaded MPE from the Polish data provider PSE S.A. [3]. Due to a lack of ground true values, we decided to use economic impact as the best metric in comparing various battery policies. It is computed as a sum of sold energy multiplied by MPE, from which the sum of bought energy multiplied by constant price regarding polish rules is subtracted.

To ensure a wider view of the problem we have also computed the total amount and value of the energy used by the considered household. It consumed 3662.48 kWh worth 4028.73 PLN in the considered period.

In Table 1 we portray the operating characteristics of select BMS models obtained as a result of running a simulation on half-a-year of obtained data. Models have been ordered according to the model's economic impact. We can observe that the worst model is On-Grid. A characteristic element of this model is the largest exchange of energy with the grid. It exported 2198.99 kWh and imported 2763.84 kWh. The rest of the models equipped with a battery had a much lower rate of electricity exchange. Due to the uncontrollability of On-Grid model it achieved the lowest averaged sell price, equal to 0.84 PLN. The total balance is -1173.70 PLN. A negative balance means that the installation owner has to purchase additional energy from the grid, but of course, the amount is much lower than if no PV would be installed.

As expected StBMS turned out to be better than On-Grid. However, it is also the worst storage-ready solution. This model generated the lowest energy exchange, but due to the behavior, it sells energy when it was the cheapest. It results in an average sell price equal to 0.88 PLN which is only 0.04 PLN better than for the On-Grid installation. Also, the final balance was only 201,78 PLN better.

The second best model was TbBMS. On one hand, it reduced energy exchange, on the other it sold more energy than StBMS. Unfortunately, also electricity import was the highest of all the battery solutions. It resulted in a good average sell price of 0.96 PLN and a balance of -865.50 PLN, which is 308,20 PLN better than On-Grid.

The best model was SBMS, which is a variation of TbBMS, but with delayed charging functionality disabled (*enable_delayed_charging* set to false). This model generated the best economic balance of -807.26 PLN, which is 366,44 PLN better than the simplest On-Grid system. This model also generated the highest average sell price at the level of 0.99 PLN.

5 Conclusions

During the research, we defined the weaknesses of state-of-the-art Battery Management System Self-Consumption. Meeting them we proposed another BMS, which makes the behavior better for the Polish system. We used economic impact as a metric to compare all considered BMSs. We found out that StBMS is better than the default On-Grid system and we proved that our models are better than state-of-the-art StBMS. We proposed two variations of the same model TbBMS, SBMS. Unexpectedly, the simpler SBMS turned out to be the best one. We obtained balances -1173.70, -971.92, -865.50 and -807.26 PLN respectively. We suppose that happened because of the selection of the forecasting model, which was Identity. The forecasts are fundamental for our models TbBMS, SBMS, however SBMS did not use PV forecasts. It only uses consumption forecasts. Taking the above into consideration we can attest that PV production is too complex to be fairly forecasted by Identity model. In further research more sophisticated forecasting models, such as ARMA, SARIMA or ANN-based [1, 5] models should be used for this purpose. Additionally, it would be better to use some kind of optimization algorithms such as Greedy or Genetic Algorithm instead of charging or discharging based only on the cheapest or the most expensive price occurrence time.

References

1. Hassanzadeh, M., Etezadi-Amoli, M., Fadali, M.: Practical approach for sub-hourly and hourly prediction of PV power output. In: North American Power Symposium 2010, pp. 1–5 (2010)
2. Pasqui, M., et al.: A new smart batteries management for renewable energy communities. Sustain. Energy Grids Netw. **34**, 101,043 (2023)
3. PSE SA: Market energy prices (2023). https://www.pse.pl/dane-systemowe/funkcjonowanie-rb/raporty-dobowe-z-funkcjonowania-rb/podstawowe-wskazniki-cenowe-i-kosztowe/rynkowa-cena-energii-elektrycznej-rce
4. Sejm of Republic of Poland: Act of 29 October 2021 on changing the Act of Renewable Energy Sources Law. (Dz.U. 2021, item. 2376, with amendments) (2021)
5. Vagropoulos, S.I., Chouliaras, G.I., Kardakos, E.G., Simoglou, C.K., Bakirtzis, A.G.: Comparison of SARIMAX, SARIMA, modified SARIMA and ANN-based models for short-term PV generation forecasting. In: 2016 IEEE International Energy Conference (ENERGYCON), pp. 1–6 (2016)

Document Annotation Tool for News Content Analysis

Marta Gackowska, Gracjan Katek, Mścisław Śrutek, Rafał Kozik, and Michał Choraś

Abstract Every day, the average Internet user perceives an abundance of content that is unintentionally consumed every day. We frequently hear the seemingly obvious remark that the modern world is full of data. We are bombarded with numerous links to amusing content circulated by our friends, various news and content providers, and social media. Unfortunately, an increasing amount of this information is only loosely related to the truth. Some of the low-quality content news could be automatically detected by modern large language models (LLM). Unfortunately, we need a large number of annotated articles to train such models. In this paper, we described our tool for news content annotation. In particular, we explain our research methodology, the tool architecture, and the analysis of the quality of the annotations. In our experiments, we engaged more than 100 volunteers, who annotated almost 4000 articles.

1 Introduction

We absorb hundreds of new pieces of information from the outside world every day, which exhausts us and renders us numb to stimuli. This is the ideal time to deceive us with low-credibility content that grabs our attention and looks appealing. Unfortunately, we are unable to independently verify all of the material we find every day. Moreover, we often have a natural tendency to take in information that seems interesting or relevant. Most of the time, we are unaware that we are republishing low-quality content and sometimes contribute to the misinformation problem.

In the literature, we may find various solutions for detecting so-called fake news [4–6, 9, 10]. These often are benchmarked on various fake news datasets (e.g. [1, 7, 8]) which concerns politics, health, COVID, presidential election, etc. Unfortunately, these datasets are mostly limited to the English language and provide labelling of the entire document or clime with limited insight into the actual content. In the literature there are also multi-modal approaches to content analysis (e.g. [2, 3]). However, in

M. Gackowska · G. Katek · M. Śrutek · R. Kozik (✉) · M. Choraś
Bydgoszcz University of Science and Technology, Al. prof. S. Kaliskiego 7, 85-796 Bydgoszcz, Poland
e-mail: rkozik@pbs.edu.pl

R. Burduk et al. (eds.), *Progress on Pattern Classification, Image Processing and Communications*, Lecture Notes in Networks and Systems 766,
https://doi.org/10.1007/978-3-031-41630-9_21

this researcher we only focus on the text. This is motivated by the fact, that our future goal is to train ML models to detect particular type of textual content (e.g. satire, political promises, or personal opinions).

Therefore in this paper, we have proposed our tool used to annotate the content of articles written in the Polish language. In the research, we engaged more than 100 volunteers, who contributed to almost 4000 annotations.

This paper is structured as follows. First, we explain the architecture of the proposed tool used for collecting annotations, Next, we demonstrate our analysis related to the quality of the obtained annotations. Finally, we conclude the paper with remarks and explain plans for future work.

2 Proposed Tool for Content Annotation

In this section, we explain our tool for the article's content annotation. The general architecture has been presented in Fig. 1. First, the database (here we used MongoDB[1] is populated with news content. In this research, we rely only on the text and we discard any other digital content (e.g. videos and images). In order to collect textual data we used Scrapy framework[2] In order to start the crawling process we provided so call seeds (e.g. URL addresses of the content providers) and let the crawler visit any other links encountered in the document bodies.

Each document in the database is assigned a unique identifier, which is based on the URL address and the date when the document has been acquired by the crawler.

The documents were assigned to the annotator randomly by a web server. In order to have better coverage of annotated documents the algorithm tended to serve documents that have not been annotated yet.

In the database, we have collected news published in the period ranging from 2021/12/31 to 2022/04/22. The rough distribution of documents per week has been shown in Fig. 2. It can be noticed that the visible peak can be observed in the period ranging from 21/02/2022 to 27/02/2022. This is the week when the Russian invasion of Ukraine happened.

In Fig. 3 we have presented the heatmap of popular topics spanning over the considered time period. It can be noticed that the news in the period ranging from 21/02/2022 to 27/02/2022 have been mostly dominated by the War in Ukraine.

In Fig. 4 we have presented the documents counted per topic. The name of the topic is assigned to the document by the publisher itself and it comes from the metadata collected by the crawler. The figure clearly shows that the news is dominated by the War in Ukraine.

[1] MongoDB homepage: https://www.mongodb.com/.

[2] Scrapy homepage: https://scrapy.org/.

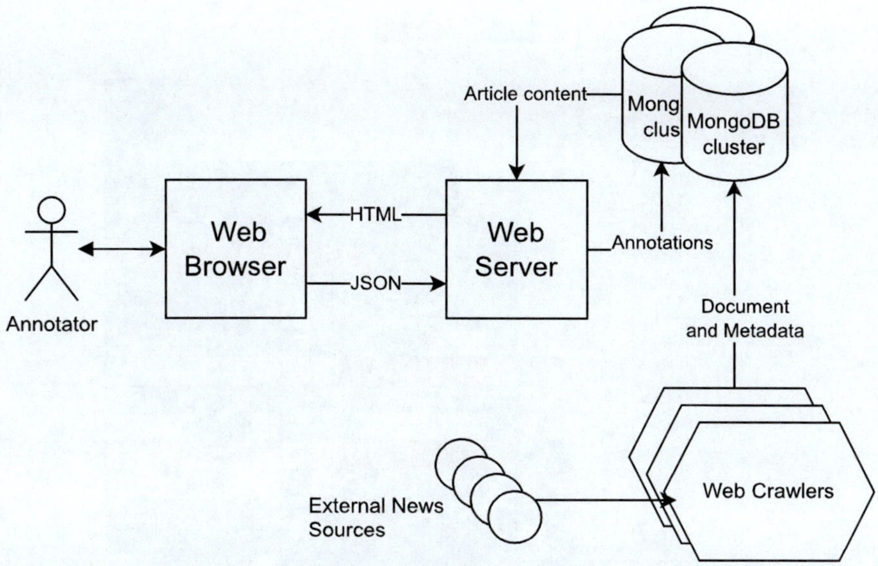

Fig. 1 The architecture of the proposed annotation tool

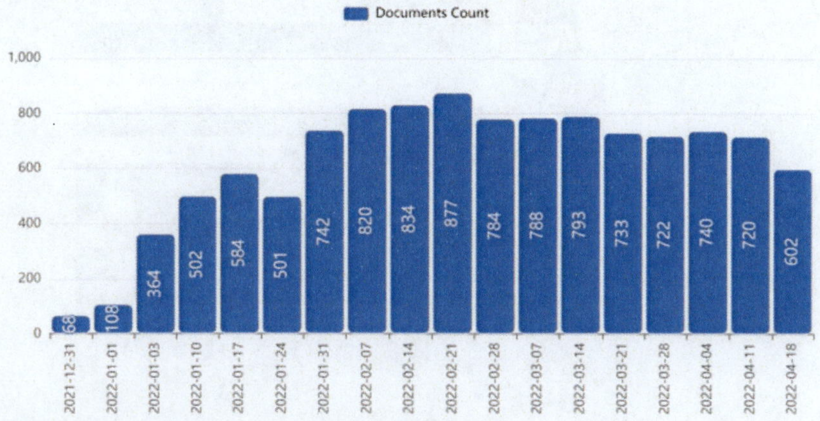

Fig. 2 The number of documents published/crawled per week.

3 Results Analysis

In this section, we briefly analysed the results obtained from the annotation process. As it is shown in Fig. 5 the annotation started in May 2022 and lasted until February 2023. From the figure itself, it is visible that there have been two phases of the annotation process. The first happened during Spring and Summer in 2022. The second phase has been started in Autumn.

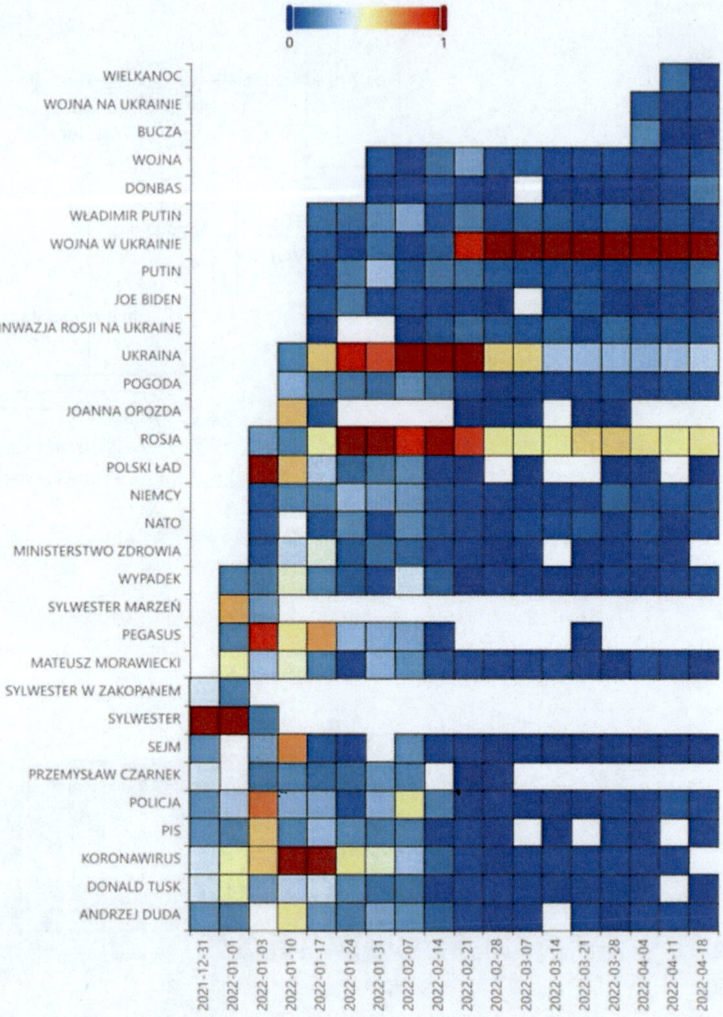

Fig. 3 Heatmap of trending topics (y-axis) over the considered time span (x-axis).

In the first phase of annotation, a significant spike in June is visible. This is due to the deadline that has been set for the first round of annotations. In that case, we may expect a low quality of annotations, because the user tended to focus on the number of annotated documents rather than on the actual content of the annotated document.

In order to measure that phenomenon we anticipated that there should be a correlation between the length of the document and the time spent on reading the document.

Inspecting the distribution of document lengths (counted in characters) shown in Fig. 6 one may notice that a significant number of documents are shorter than 1500

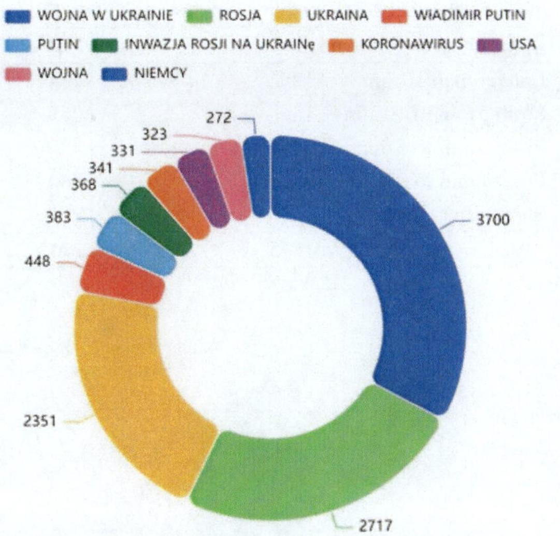

Fig. 4 The number of documents per topic (identified/assigned by the publisher)

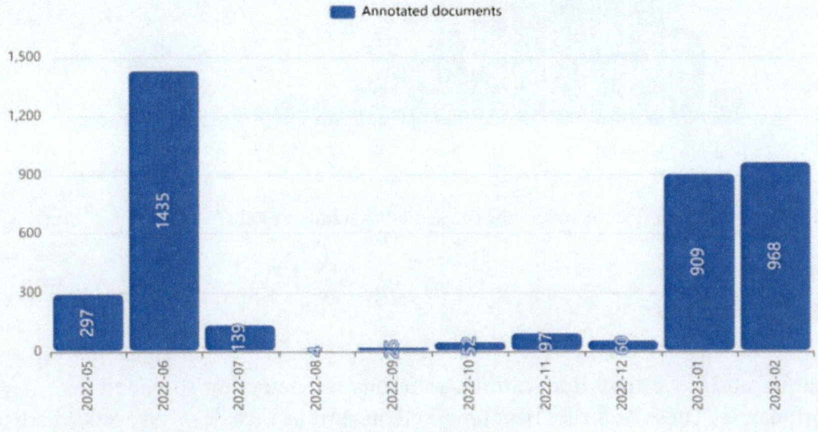

Fig. 5 The number of documents annotated per month

characters. In that regard, a rough calculation indicates that the user shall spend (on average) more than 1 min to annotate the document. In that regard, we have calculated the number of documents falling into a specific range of time spent by the annotator to read the document. In Table 1, one may see that a quite big number of documents have been annotated in less than 1 min. This issue is visible for both of the phases.

During the process of ML model training, we will have to take this issue into consideration, because the low quality of the annotation will certainly deteriorate the model classification performance.

Table 1 Number of documents with respect to the duration of the annotation process

Duration of annotation	Phase 1	Phase 2	Total
Longer than 10 min	185	164	349
From 5 min to 10 min	139	259	398
From 2 min to 5 min	376	433	809
From 1 min to 2 min	392	484	876
Shorter than 1 min	783	771	1554
	1875	2111	3986

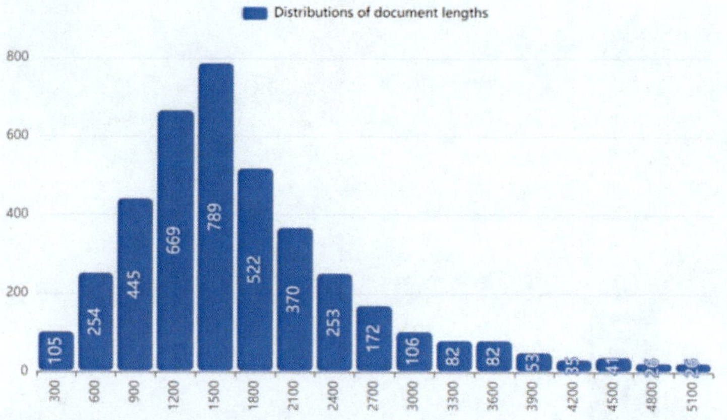

Fig. 6 Distributions of document lengths (measured as number of characters)

4 Conclusions

In the literature, we may find various solutions for detecting so-called fake news. Unfortunately, These often are benchmarked on various fake news datasets which are mostly limited to the English language and provide labelling of the entire document or clime with limited insight into the actual content.

Therefore in this paper, we have proposed our tool used to annotate the content of articles written in the Polish language. In the research, we asked more than 100 volunteers to annotate news articles. As a result, these activities resulted in almost 4000 annotations.

In the experiment section, we made an attempt to assess the quality of the annotations obtained in the two phases. In particular, we have correlated the time spent by a user on annotating the document with the length of the content (measured in the number of characters). The analysis revealed that there are indeed several annotations (in both of the phases) that may have questionable quality.

Our plan for future work will be to mitigate this issue and encourage the annotator to focus on more the annotation process rather than on the number of annotations.

Moreover, we plan to repeat the annotation of the same document in order to contradict different opinions expressed by the annotators.

References

1. Banik, S.: Covid fake news dataset [data set]. Zenodo, Online (2020). https://doi.org/10.5281/zenodo.428252
2. Bazmi, P., Asadpour, M., Shakery, A.: Multi-view co-attention network for fake news detection by modeling topic-specific user and news source credibility. Inf. Process. Manag. **60**(1), 103146 (2023). https://doi.org/10.1016/j.ipm.2022.103146. https://www.sciencedirect.com/science/article/pii/S0306457322002473
3. Jing, J., Wu, H., Sun, J., Fang, X., Zhang, H.: Multimodal fake news detection via progressive fusion networks. Inf. Process. Manag. **60**(1), 103120 (2023). https://doi.org/10.1016/j.ipm.2022.103120. https://www.sciencedirect.com/science/article/pii/S0306457322002217
4. Khan, J.Y., Khondaker, M.T.I., Afroz, S., Uddin, G., Iqbal, A.: A benchmark study of machine learning models for online fake news detection. Mach. Learn. Appl. **4**, 100032 (2021). https://doi.org/10.1016/j.mlwa.2021.100032. https://www.sciencedirect.com/science/article/pii/S266682702100013X
5. Ksieniewicz, P., Zyblewski, P., Borek-Marciniec, W., Kozik, R., Choraś, M., Woźniak, M.: Alphabet flatting as a variant of n-gram feature extraction method in ensemble classification of fake news. Eng. Appl. Artif. Intell. **120**, 105882 (2023). https://doi.org/10.1016/j.engappai.2023.105882
6. Kula, S., Kozik, R., Choraś, M.: Implementation of the BERT-derived architectures to tackle disinformation challenges. Neural Comput. Appl. (2021). https://doi.org/10.1007/s00521-021-06276-0
7. Li, Y., Jiang, B., Shu, K., Liu, H.: MM-COVID: a multilingual and multimodal data repository for combating COVID-19 disinformation (2020)
8. Risdal, M.: Getting real about fake news. Kaggle, Online (2016). https://www.kaggle.com/mrisdal/fake-news
9. Szczepański, M., Pawlicki, M., Kozik, R., Choraś, M.: New explainability method for BERT-based model in fake news detection. Sci. Rep. **11**(1), 23705 (2021). https://doi.org/10.1038/s41598-021-03100-6
10. Zhang, J., Dong, B., Yu, P.S.: Fakedetector: effective fake news detection with deep diffusive neural network. In: 2020 IEEE 36th International Conference on Data Engineering (ICDE), pp. 1826–1829. IEEE Computer Society, Los Alamitos (2020). https://doi.org/10.1109/ICDE48307.2020.00180. https://doi.ieeecomputersociety.org/10.1109/ICDE48307.2020.00180

Dynamic Time Warping Technique Applied to the User's Intent Recognition for Myoelectric-Based Control of Upper Limb Prosthesis

Pawel Trajdos and Marek Kurzynski

Abstract Recognition of the surface myoelectric (sEMG) signal to decode the user's intention is the most advanced approach to the control of a bionic upper limb prosthesis. This paper is devoted to the methods of distance-based recognition of the sEMG signals using the dynamic time warping (DTW) technique to determine the distance measures between the sEMG signals. Three methods are proposed and developed: a method in which distances are directly used in KNN classifier and two methods in which KNN classifier is fed with features extracted from distances using multidimensional scaling technique, and clustering-based approach, both methods in two versions. The proposed methods were experimentally compared for different K in the KNN classifier and for different quality measures. In addition, for methods with features extracted from distances, the quality of classification for different models of classifiers was experimentally tested. Experimental investigations were conducted using real sEMG signals coming from an amputee and from an able-body subject with an immobilised hand simulating an amputation.

Keywords Classification · Myoelectric hand prosthesis control · Dynamic time warping

1 Introduction

The hand is one of the most complex organs of the human body. The hand is not only an excellent manipulator but also, through numerous receptors, a source of complex tactile sensations. The loss of a hand dramatically worsens the quality of a person's life. Loss of both hands makes independent functioning practically impossible. In such cases, equipping the amputee with the so-called bionic hand prosthesis, i.e.,

P. Trajdos (✉) · M. Kurzynski
Wroclaw University of Science and Technology, Wroclaw, Poland
e-mail: pawel.trajdos@pwr.edu.pl

M. Kurzynski
e-mail: marek.kurzynski@pwr.edu.pl

R. Burduk et al. (eds.), *Progress on Pattern Classification, Image Processing and Communications*, Lecture Notes in Networks and Systems 766,
https://doi.org/10.1007/978-3-031-41630-9_22

a prosthesis controlled by biological signals, can be a significant support. Modern bioprostheses are typically controlled by surface electromyographic (sEMG) signals from residual limb muscles, and their mechanical design is fully anthropomorphic, almost approximating the dexterity of the human hand [6].

As a result of several decades of development of methods and tools for myoelectric control of hand bioprosthesis, three approaches have emerged [26]:

1. Non-recognition control [9]. This framework covers: (1) on-off control – if the sEMG signal exceeds a threshold, a certain prosthesis function is activated; (2) proportional control which permits to vary force and speed of movement proportionally to the amplitude of the recorded sEMG signal; (3) finite state machine – its states represent pre-defined prosthesis movements and transition rules are associated with sEMG signal.
2. Recognition-based control. Myoelectric pattern recognition to decode the intent of the user is the most advanced approach to the control of a powered bioprosthesis. In this scheme, the desired classes of movements (grips or manipulations) are discriminated on the basis of sEMG signal patterns by the recognition system, and the variety of prosthesis functions depends directly on the classification performance [7].
3. Regression-based control [16]. In the regression technique, a continuous output value is estimated for each degree of freedom (DoF). Such control allows natural, smooth and simultaneous movement of multiple DoFs of prosthesis.

Since the topic of this paper falls under the second approach, we will start with a brief look at the recognition-based control system shown in Fig. 1. First, in the human controller, the intention of the prosthesis to perform a particular movement arises in the user's mind. This intention is converted in the motor cortex into a neural signal that, when transmitted to the forearm, activates the residual muscles of the stump. The activity of these muscles generates an EMG biosignal that is transferred to the bioprosthesis (control plant) as a control variable. The purpose of the measurement system is to record the sEMG signal generated by a human and further processed in the prosthesis (after A/D conversion), therefore it can be called the human-prosthesis (sEMG) interface. This interface should have several features that facilitate stable and repeatable sEMG recording [13].

In the recognition-based control paradigm, the main part of the prosthesis software is a complex decision-making system called a recogniser or recognition system.

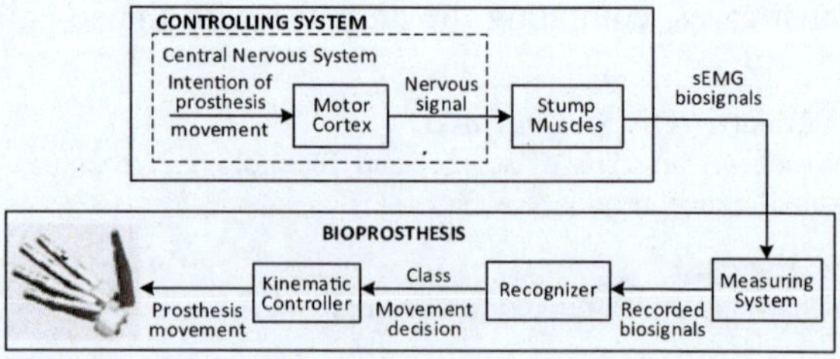

Fig. 1 Block-diagram of the recognition-based system for myoelectric control of upper limb prosthesis.

Its typical structure shown in Fig. 2A includes the following stages of processing and analysis of the recorded sEMG signal: (1) feature extraction (e.g. using Fast Fourier Transform (FFT), Discrete Wavelet Transform (DWT) or Autoregressive Model (AR)) [27]; (2) feature selection (via filters, wrappers, or embedded methods) or feature reduction (e.g. PCA method) [37], and (3) classification (different classifier models [26] and different classification schemes [24, 25]). We assume that the classifier is built on the basis of the supervised learning method, therefore Fig. 2A also shows (a dashed line) a diagram of the processing of the training sEMG signals in the classifier learning phase.

Finally, the task of the kinematic controller is to control the degrees of freedom (DoFs) of the prosthesis in such a way that the movement of the mechanical structure is consistent with the classification result or – if the classification is correct – with the user's intention.

In the paper, the authors propose new schemes of the recognition system intended for sEMG-based control of the upper limb prosthesis. A different method of operation of the recognition system is the result of the use of the dynamic time warping (DTW) technique in sEMG signal processing.

DTW is one of the popular methods applied in time-series data mining [12]. It is used to determine a distance measure between two time series. Unlike lock-step measures, where the i-th point of one series is compared with the i-th point of another (e.g. Euclidean distance), the DTW technique produces a flexible measure that creates nonlinear mapping in order to align the series and allow comparison of one-to-many points. This allows us to capture different types of dissimilarities of compared time-series, e.g. the shape of the series [1].

The DTW technique plays a particularly important role in distance-based time-series recognition. According to the taxonomy presented in [1], the methods can be divided into three main categories that differ in the way the distances determined by the DTW technique are used. These distances: (1) are directly used in a minimum-distance classifier; (2) are used to obtain the distance-based features, which are then applied to a plain classifier; (3) are used to obtain kernels for time series.

Distance-based time series recognition methods are widely applied in various practical tasks. They also include recognition of biosignals, in particular, sEMG signals. As examples can be mentioned: handwritting recognition [31], recognition of hand gesture [33], EEG- and EMG-based evaluation of movement functional rehabilitation after stroke [20], authentication system [23], sign language handshape recognition [3], silent speech recognition [14], a fatigue assessment [10], finger gesture recognition [2], analysis of human-computer interactions (wake words) [22].

The paper [21] presents the use of a distance-based recognition system with the DTW technique to sEMG-based control of a hand bioprosthesis. The authors limited themselves to the use of the one method of building a distance-based recognition system, and sEMG signals from able-bodied persons were used in the study. In our research, we investigate two approaches to building distance-based recognition and we use sEMG signals from an amputee and from an immobilised hand simulating an amputation. Such sEMG signals are much more difficult to analyse than signals from able-bodied persons because they are less discriminating between classes and, moreover, they are non-stationary. The first circumstance results from the fact that the class (label) of the signal is not caused by the real movement of the hand, but is caused by the movement of the hand that does not exist (the phantom hand) or "movement without movement" (the immobilised hand), and this situation results in a weaker correlation between the class and the sEMG signal [36]. The non-stationarity of sEMG signals results from the subject's fatigue and changing recording conditions (movements of the prosthesis, sensors shifting). Thus, the cognitive purpose of the work is related to the assessment of the usefulness of the distance-based recognition system in the sEMG control of the bioprosthetic hand and the comparative experimental analysis of the methods tested in a real environment.

The main goal of the paper (and the conducted experimental study) is to answer the following research questions: **RQ1**: When considering the usage of DTW as a distance measure between time series, does the usage of distance-based or embedded features offer significantly different classification quality when compared to the use of a raw distance-based approach? **RQ2**: Does the usage of a classifier other than the Nearest-Neighbours classifier for the features extracted using DTW causes any change in the obtained classification quality? **RQ3**: What is the impact of averaging channel-specific distance matrices on classification quality?

The paper is divided into five sections and is organised as follows. In Sect. 2 classification models applied in experiments are presented and three schemes of distance-based recognition systems with the DTW technique are developed. Section 3 describes in detail two sets of sEMG signals used in experiments and provides an insight into experimental setup. The results of experimental studies with discussion are presented in Sect. 4. Section 5 concludes the paper.

2 Methods

As mentioned in the previous section, DTW is mainly used directly as a method of calculating between-time series distances. These distances are used by minimal-distance recognition methods, such as the K Nearest-Neighbours (K-NN) classifier. For a new-coming sEMG signal, DTW is used to calculate the distances to the training objects, and then the prediction is made. This approach is visualised in Fig. 2B. The strategy is, however, limited to distance-based classifiers. To use different types of classifiers, another approach must be used. These alternative approaches transform distances into a vector of attributes that can be fed to any classification algorithm [1]. A general schema of such approaches is depicted in Fig. 2C. This group may be further divided into three subgroups of feature extractors: global feature extractors, local feature extractors, and embedded ones [1].

The methods in the first group produce feature vectors using distances to previously stored time series. The main concern is which time series should be stored to produce a new feature vector [38]. Generally, the choice may be random or data-driven. In our experimental study, we decided to use one approach from this group. The used method selects the sequences to store using the agglomerative clus-

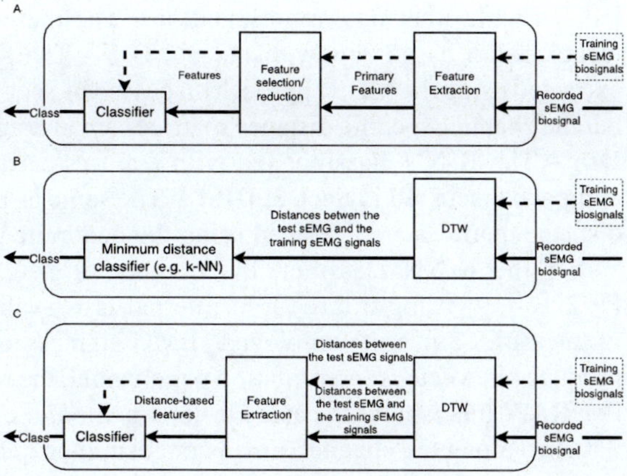

Fig. 2 Recognition system structures

tering approach [32]. That is, the data are clustered with the number of clusters selected using the average silhouette criterion [28]. The choice is made within the set $\{1, 2, \cdots, \lceil\sqrt{N}\rceil\}$, where N is the number of objects in the training set. This clustering is performed for each channel of the sEMG signal. Then, for each cluster, the cluster representatives are selected using also the silhouette criterion.

The second group of feature extractors is somewhat similar to the first one. However, instead of storing entire sequences, the method stores only fragments of time series. An example of such a method is the so-called shapelet approach [15]. Methods belonging to this group are not used during the experimental study performed.

The third group of feature extractors produces so-called embedded features. In contrast to the previously mentioned groups, these methods do not use the distances directly as the new representation (feature vector) [1]. Instead, they use the distances indirectly to produce a new representation. A well-known example of this approach is the Multidimensional Scaling (MDS) technique [34]. This approach utilises a pairwise distance matrix to create a vector representation that produces an approximate version of the original distance matrix. When the new representation for the training set vectors is known, then a representation of a new time series may be obtained by solving an optimisation problem that binds training time series and their representation [34]. In the experimental study conducted, an evolutionary algorithm is used to solve this optimisation problem [4]. When the MDS technique is used, an issue of dimensionality selection for the new representation must also be addressed. One of the possible solutions is to use knee/elbow detection in the stress function plot [29].

3 Experimental Setup

To answer the research questions presented in Sect. 1, two experiments are proposed.

The first compares the classification quality of the raw distance-based Nearest-Neighbour classifier with the classification quality obtained by the same classifier but fed with other data representations. The methods compared: **RAW** – The K-NN classifier based on raw distance; **MDS** – The K-NN classifier fed with features extracted using the MDS approach discussed in Sect. 2; **MDSf** – The same as previous but the channel-specific distance matrices are averaged using the quadratic mean [8]; **DIS** – The K-NN classifier fed with features extracted using the clustering-based approach discussed in Sect. 2; **DISf** – The same as previous but the channel-specific distance matrices are averaged using the quadratic mean.

For the K-NN classifier, the number of neighbours is chosen from the set $\{1, 2, \cdots, 9\}$ using a three-fold internal cross-validation and accuracy as a quality measure. Since the employed sEMG signals are multi-channel **MDS**, and **DIS** produce one vector representation per channel, the representations are concatenated. For **RAW** this is not the case. For this method, the multichannel distance matrix is flattened along the channel dimension, using the quadratic mean [8]. In other words, channel-specific distance matrices are averaged. This is due to the fact that the K-NN classifier (in its basic version) can use only one distance matrix [19]. Another solution may have been to use an ensemble of K-NN classifiers, but it would result in a comparison between a single K-NN classifier for **MDS**, and **DIS** and a classifier committee for **RAW**. Therefore, we decided to use the averaged distance matrix for the **RAW** classifier. However, to properly assess the impact of channel-specific distance matrices averaging, we also employed **MDSf** and **DISf** extractors.

In the second experiment, we use only the representations obtained using **MDS**, **MDSf**, **DIS**, and **DISf** approaches. For each extraction technique, we compare the classification quality obtained by the following classifiers: Random Forest classifier (**RF**) with the committee size set to 20 [5]; Naive Bayes classifier (**NB**) employing Gaussian distribution for modelling the class-conditional distributions [17]; **K-NN** classifier with the number of neighbours is chosen from the set $\{1, 2, \cdots, 9\}$ using the three-fold internal cross-validation and accuracy as a quality measure [19].

In both experimental studies, classification quality is assessed using the following quality criteria [30]: accuracy, F_1-measure, precision, and recall. The training and testing sets are extracted using stratified ten-fold crossvalidation.

Following the recommendations of [11], the statistical significance of the results obtained was evaluated using pairwise tests using the Wilcoxon signed-rank test [11]. To control family-wise errors (FWER) of the Wilcoxon test procedure, the Holm approach was employed [18]. For all tests, the significance level was set at $\alpha = .05$.

Two sets of sEMG signals were used in experimental investigations:

- **Dataset 1**. Male able-body subject, aged 73, right-handed. In creating the set of signals, a simulation of amputation was modelled by completely immobilising the hand during the recording of signals, which obliges the subject to generate EMG signals based on his own image of the movement performed, and not on the basis of actual movements performed by the hand. The sEMG biosignals were registered using a special designated measuring circuit with 8 sensors equally spaced located around the forearm and a sampling frequency of 500 Hz [35]. The dataset used in the experiments consisted of eight classes of movement images and 100 eight-channel sEMG signals for each class. Each measurement lasted 1000 ms and was preceded with a 3 s break.

– **Dataset 2**. Female amputee, 46 years old, right-handed, with traumatic deficiency below the elbow on the left side. The measuring circuit, the number of sensors (channels), and a sampling frequency as in the first dataset. The set of signals used in the experiments consisted of five classes of movement intent and 60 eight-channel sEMG signals for each class. Each measurement lasted 1000 ms and was preceded with a 5 s break.

Both subjects gave informed consent prior to participation.

4 Results and Discussion

This section presents the results of the experimental study accompanied by a statistical analysis and discussion of the results. The section is divided into two subsections, one for each of the conducted experiments.

4.1 Experiment 1

This experiment compares the classification quality of the K-NN algorithm that uses a raw distance matrix against the K-NN classifier fed using extracted features. Classification quality results (for both sets) are presented in Table 2. Each row of the table is related to one method and each column is related to classification criterion. Each entry in the table presents the average result across all ten folds and the standard deviation. The bolded entries are related to the best result in the column. The results of the statistical evaluation are presented in Tables 2 and 3. The tables presents p-values for the pairwise Wilcoxon tests. The bolded entries mean that there is a significant difference between the two methods.

Let us begin with an analysis of the results for set no. 1. It is clear that, for all quality criteria, the **RAW** method is the worst performing one, whereas the **DIS** method is the best one. The differences in the average values of the quality measures are substantial and consequently the statistical test confirms their significance. However, when we use the channel-averaged distance matrix to the **DIS** method (approach denoted as **DISf**), the quality loss is significant in terms of all quality criteria. It may suggest that the source of discriminatory information lies in the pattern differences between the sEMG channels. This observation is also confirmed by the results obtained for **MDS** and **MDSf** methods. For this pair of methods, a quality drop after applying distance matrix averaging is also observable. However, the difference is not always statistically significant. But still **RAW** performs worse than **DISf** and **MDSf**, so distance matrix averaging is not only one reason for this. The other reason may be related to the fact that the **DIS** and **MDS** methods lose some information compared to **RAW**, which

Table 1 Classification quality for datasets 1 and 2.

	ACC	F1	precision	recall
RAW	.376 ± .032	.299 ± .028	.383 ± .067	.378 ± .032
MDS	.761 ± .029	.759 ± .032	.789 ± .034	.764 ± .029
MDSf	.733 ± .039	.729 ± .041	.745 ± .031	.736 ± .040
DIS	**.958** ± .011	**.958** ± .011	**.962** ± .009	**.959** ± .011
DISf	.559 ± .043	.541 ± .041	.546 ± .043	.562 ± .043

(a) Dataset 1

	ACC	F1	precision	recall
RAW	.313 ± .070	.275 ± .082	.350 ± .141	.313 ± .070
MDS	.473 ± .100	.457 ± .100	.477 ± .108	.473 ± .100
MDSf	.363 ± .048	.351 ± .055	.360 ± .069	.363 ± .048
DIS	**.513** ± .056	**.511** ± .059	**.573** ± .087	**.513** ± .056
DISf	.257 ± .087	.260 ± .090	.292 ± .114	.257 ± .087

(b) Dataset 2

Table 2 Statistical analysis for dataset 1.

	ACC				F1				precision				recall			
	MDS	MDSf	DIS	DISf	MDS	MDSf	DIS	DISf	MDS	MDSf	DIS	DISf	MDS	MDSf	DIS	DISf
RAW	**.004**	**.004**	**.004**	**.004**	**.004**	**.004**	**.004**	**.004**	**.004**	**.004**	**.004**	**.004**	**.004**	**.004**	**.004**	**.004**
MDS	1.00	.275	**.004**	**.004**	1.00	.232	**.004**	**.004**	1.00	**.027**	**.004**	**.004**	1.00	.232	**.004**	**.004**
MDSf		1.00	**.004**	**.004**		1.00	**.004**	**.004**		1.00	**.004**	**.004**		1.00	**.004**	**.004**
DIS			1.00	**.004**			1.00	**.004**			1.00	**.004**			1.00	**.004**

Table 3 Statistical analysis for dataset 2.

	ACC				F1				precision				recall			
	MDS	MDSf	DIS	DISf	MDS	MDSf	DIS	DISf	MDS	MDSf	DIS	DISf	MDS	MDSf	DIS	DISf
RAW	.100	.284	**.043**	.284	**.049**	.193	**.023**	.922	.422	.625	.117	.625	.069	.343	**.042**	.343
MDS	1.00	.054	.284	**.016**	1.00	**.049**	.387	**.016**	1.00	.068	.625	.068	1.00	**.035**	.343	**.016**
MDSf		1.00	**.016**	.083		1.00	**.016**	.129		1.00	**.018**	.580		1.00	**.016**	.084
DIS			1.00	**.016**			1.00	**.016**			1.00	**.018**			1.00	**.016**

uses all the information encoded in the distance matrix. The **MDS** method loses some information during the construction of the low-dimensional representation of the distance matrix. Similarly, the **DIS** method does not use all points from the training set, but it uses only a subset selected by the clustering algorithm. This small information loss is not harmful, but it increases the generalisation ability of the classifier fed by the extracted data.

For the dataset no. 2, the classification quality also drops when the distance matrix averaging is applied. However, not all observed differences are statistically significant. In addition, the **DISf** and **MDSf** methods do not differ significantly from the **RAW** method. It may suggest that, for patients with amputated limbs, cross-channel information is even more important than for able-bodied patients. This observation may also be explained by the fact that the patient, from whom the sEMG signals were collected, did not have phantom limb sensation. Consequently, she may have been tried to generate different class patterns by differentiating muscle tension in different parts of the stump (Table 1).

Table 4 Classification quality for dataset 1.

	ACC, MDS	ACC, DIS	F1, MDS	F1, DIS	precision, MDS	precision, DIS	recall, MDS	recall, DIS
RF	.874 ± .034	**.958 ± .018**	.874 ± .034	**.958 ± .018**	.880 ± .036	.962 ± .016	.875 ± .034	.958 ± .018
NB	**.922 ± .023**	.876 ± .030	**.922 ± .023**	.870 ± .033	**.933 ± .020**	.903 ± .034	**.925 ± .022**	.881 ± .030
KNN	.799 ± .048	**.958 ± .011**	.799 ± .046	**.958 ± .011**	.824 ± .046	**.962 ± .009**	.802 ± .048	**.959 ± .011**
RFf	.738 ± .052	.501 ± .053	.739 ± .049	.493 ± .052	.751 ± .051	.510 ± .052	.740 ± .051	.504 ± .053
NBf	.788 ± .041	.551 ± .057	.782 ± .045	.504 ± .059	.794 ± .052	.487 ± .059	.793 ± .041	.556 ± .058
KNf	.752 ± .034	.554 ± .037	.751 ± .032	.536 ± .036	.762 ± .035	.542 ± .039	.755 ± .033	.557 ± .038

Table 5 Classification quality for dataset 2.

	ACC, MDS	ACC, DIS	F1, MDS	F1, DIS	precision, MDS	precision, DIS	recall, MDS	recall, DIS
RF	**.583 ± .099**	**.853 ± .045**	**.581 ± .099**	**.852 ± .048**	**.610 ± .100**	**.874 ± .036**	**.583 ± .099**	**.853 ± .045**
NB	.563 ± .084	.523 ± .100	.550 ± .080	.510 ± .102	.581 ± .102	.587 ± .105	.563 ± .084	.523 ± .100
KNN	.417 ± .075	.513 ± .056	.402 ± .073	.511 ± .059	.450 ± .108	.573 ± .087	.417 ± .075	.513 ± .056
RFf	.357 ± .094	.270 ± .081	.354 ± .095	.259 ± .082	.399 ± .114	.263 ± .089	.357 ± .094	.270 ± .081
NBf	.317 ± .112	.290 ± .056	.290 ± .125	.229 ± .047	.328 ± .165	.283 ± .080	.317 ± .112	.290 ± .056
KNf	.373 ± .083	.257 ± .087	.361 ± .083	.260 ± .090	.395 ± .105	.292 ± .114	.373 ± .083	.257 ± .087

4.2 *Experiment 2*

This experiment compares the classification quality obtained by classifiers other than KNN using different DTW-based data representations. The average values and standard deviations of the employed quality criteria are presented in Tables 4 and 5. The results of the statistical evaluation are shown in Tables 6 and 7.

Let us begin with the analysis of the impact of applying other classifiers to the data extracted using averaged channel-specific distance matrices. As observed in the first experiment, the classification qualities are lower when distance matrix averaging is applied. This is because cross-channel information is lost. Applying other classifiers, which may use more cross-channel information, does not cause any improvement in classification quality. In addition, for the able-bodied person and the Random Forest classifier (**RFf**) applied with **DIS** extractor, the classification quality is significantly lower compared to the **KNf** classifier.

The situation changes when analysing the feature extractors that construct channel-specific features. For the dataset taken from the post-amputation limb, classifiers other than **KNN** achieve higher values for all quality criteria. More importantly, for the **DIS** extractor, the **RF** classifier significantly outperforms the **KNN** classifier. This is probably due to the fact that the Random Forest classifier may make more use of cross-channel attribute value differences than the other investigated classifiers. The situation is a bit different for the dataset obtained for the able-bodied patient. In this case, for the **MDS** extractor, the **RF** and **NB** classifiers perform significantly better than the **KNN** classifier. On the contrary, for the **DIS** extractor the **NB** classifiers perform significantly worse, and the **RF** classifier does not achieve significantly different results than **KNN**. This also confirms the earlier observation that for the patient with amputation, cross-channel differences play a more important role in class separation than in the case of the able-bodied patient.

Table 6 Statistical analysis for dataset 1.

	ACC, MDS					ACC, DIS					F1, MDS					F1, DIS				
	NB	KNN	RFf	NBf	KNf	NB	KNN	RFf	NBf	KNf	NB	KNN	RFf	NBf	KNf	NB	KNN	RFf	NBf	KNf
RF	.068	**.045**	**.018**	**.018**	**.018**	**.010**	.887	**.010**	**.010**	**.010**	**.016**	**.027**	**.016**	**.016**	**.016**	**.010**	.889	**.010**	**.010**	**.010**
NB	1.00	**.018**	**.018**	**.018**	**.018**	1.00	**.010**	**.010**	**.010**	**.010**	1.00	**.016**	**.016**	**.016**	**.016**	1.00	**.010**	**.010**	**.010**	**.010**
KNN		1.00	.101	.438	.111		1.00	**.010**	**.010**	**.010**		1.00	.137	.322	.137		1.00	**.010**	**.010**	**.010**
RFf			1.00	.152	.438			1.00	.480	**.047**			1.00	.316	.322			1.00	.889	.078
NBf				1.00	.101				1.00	.887				1.00	.252				1.00	.393
	precision, MDS					precision, DIS					recall, MDS					recall, DIS				
	NB	KNN	RFf	NBf	KNf	NB	KNN	RFf	NBf	KNf	NB	KNN	RFf	NBf	KNf	NB	KNN	RFf	NBf	KNf
RF	**.041**	.082	**.018**	**.018**	**.018**	**.010**	.889	**.010**	**.010**	**.010**	**.016**	**.035**	**.016**	**.016**	**.016**	**.010**	.779	**.010**	**.010**	**.010**
NB	1.00	**.018**	**.018**	**.018**	**.018**	1.00	**.010**	**.010**	**.010**	**.010**	1.00	**.016**	**.016**	**.016**	**.016**	1.00	**.010**	**.010**	**.010**	**.010**
KNN		1.00	.137	.465	**.027**		1.00	**.010**	**.010**	**.010**		1.00	.082	.407	.111		1.00	**.010**	**.010**	**.010**
RFf			1.00	.349	.557			1.00	.551	.316			1.00	.111	.387			1.00	.393	**.031**
NBf				1.00	.349				1.00	.109				1.00	**.046**				1.00	.779

Table 7 Statistical analysis for dataset 2.

	ACC, MDS					ACC, DIS					F1, MDS					F1, DIS				
	NB	KNN	RFf	NBf	KNf	NB	KNN	RFf	NBf	KNf	NB	KNN	RFf	NBf	KNf	NB	KNN	RFf	NBf	KNf
RF	.677	.061	.086	**.023**	**.023**	**.010**	**.010**	**.010**	**.010**	**.010**	.846	**.018**	.078	**.018**	**.018**	**.010**	**.010**	**.010**	**.010**	**.010**
NB	1.00	.059	**.023**	**.023**	.060	1.00	1.00	**.010**	**.010**	**.010**	1.00	**.018**	**.018**	**.018**	**.018**	1.00	.922	**.010**	**.010**	**.010**
KNN		1.00	.562	.348	.677		1.00	**.010**	**.010**	**.010**		1.00	.625	.191	.846		1.00	**.010**	**.010**	**.010**
RFf			1.00	.677	.677			1.00	1.00	1.00			1.00	.846	.846			1.00	.922	.922
NBf				1.00	.677				1.00	.686				1.00	.826				1.00	.922
	precision, MDS					precision, DIS					recall, MDS					recall, DIS				
	NB	KNN	RFf	NBf	KNf	NB	KNN	RFf	NBf	KNf	NB	KNN	RFf	NBf	KNf	NB	KNN	RFf	NBf	KNf
RF	1.00	**.047**	.078	**.035**	**.035**	**.010**	**.010**	**.010**	**.010**	**.010**	.695	.061	.086	**.023**	**.023**	**.010**	**.010**	**.010**	**.010**	**.010**
NB	1.00	**.035**	**.023**	**.023**	.078	1.00	.922	**.010**	**.010**	**.010**	1.00	.057	**.023**	**.023**	**.036**	1.00	.889	**.010**	**.010**	**.010**
KNN		1.00	.697	.527	.826		1.00	**.010**	**.010**	**.010**		1.00	.562	.208	.695		1.00	**.010**	**.010**	**.010**
RFf			1.00	.967	1.00			1.00	.922	.922			1.00	.695	.695			1.00	.889	.889
NBf				1.00	.967				1.00	.922				1.00	.695				1.00	.553

5 Conclusions

In this paper, we addressed the issue of building feature extraction procedures based on the dynamic time warping approach. The procedures were applied to the practical task of recognition of movement intention for amputee patients and body-abled. We conducted two experiments to provide answers to three research questions. Each of the experiments was carried out for two datasets. They provide the following answers to the research questions. (**RQ1**) Yes, the feature extraction procedure plays an important role in the task of recognising the intention of movement. Additionally, using raw distances to perform KNN classification causes the classification quality to be significantly lower. This is an important observation, since many DTW-based sEMG recognition systems use the raw distance to perform the KNN classification. (**RQ2**) The usage of classifiers other than KNN may result in improvement of classification quality. (**RQ3**) The averaging of channel-specific distance matrices has a significant negative impact on the classification quality achieved by the system. What is more, for the patient with amputation, more information about movement intent seems to be encoded in cross-channel characteristics.

The results obtained are quite interesting, but to achieve more general conclusions, the experiments must be conducted on a larger cohort of patients. In particular, we need to test more patients with amputation. This is because we have observed that the characteristics of sEMG signals recorded on a stump differ significantly from those obtained from the able-bodied patient. These differences are significant even if a limb of the able-body patient is immoblilised. This is also worth investigating other classifiers than Random Forest and Naive Bayes.

Acknowledgments This work is supported by the National Center for Research and Development (www.ncbr.gov.pl) through project no. **/0018/2020-00** (*"Things are for people"* programme). The authors have no conflict of interest to declare.
The authors thank Dr. Andrzej Wolczowski for valuable discussions, Dr. Jerzy Witkowski for the design and manufacture of the EMG/MMG sensors and Dr. Michal Bledowski for the implementation of the application for the recording of the EMG signal [35].

References

1. Abanda, A., Mori, U., Lozano, J.A.: A review on distance based time series classification. Data Min. Knowl. Disc. **33**(2), 378–412 (2018)
2. Amma, C., Krings, T., Böer, J., Schultz, T.: Advancing muscle-computer interfaces with high-density electromyography. In: Proceedings of the 33rd Annual ACM Conference on Human Factors in Computing Systems. ACM (2015)
3. Amor, A.B.H., Ghoul, O.E., Jemni, M.: Sign language handshape recognition using Myo armband. In: 2019 7th International Conference on ICT & Accessibility (ICTA). IEEE (2019)
4. Bartz-Beielstein, T., Branke, J., Mehnen, J., Mersmann, O.: Evolutionary algorithms. Wiley Interdiscip. Rev. Data Mining Knowl. Discov. **4**(3), 178–195 (2014)
5. Breiman, L.: Random forest. Mach. Learn. **45**(1), 5–32 (2001)
6. Calado, A., Soares, F., Matos, D.: A review on commercially available anthropomorphic myoelectric prosthetic hands, pattern-recognition-based microcontrollers and sEMG sensors used for prosthetic control. In: 2019 IEEE International Conference on Autonomous Robot Systems and Competitions (ICARSC). IEEE (2019)
7. Campbell, E., Phinyomark, A., Scheme, E.: Current trends and confounding factors in myoelectric control: limb position and contraction intensity. Sensors **20**(6), 1613 (2020)
8. de Carvalho, M.: Mean, what do you mean? Am. Stat. **70**(3), 270–274 (2016)
9. Ciancio, A.L., et al.: Control of prosthetic hands via the peripheral nervous system. Front. Neurosci. **10** (2016)
10. Dang, Y., Liu, Z., Yang, X., Ge, L., Miao, S.: A fatigue assessment method based on attention mechanism and surface electromyography. Internet Things Cyber-Phys. Syst. **3**, 112–120 (2023)
11. Demšar, J.: Statistical comparisons of classifiers over multiple data sets. J. Mach. Learn. Res. **7**, 1–30 (2006)
12. Esling, P., Agon, C.: Time-series data mining. ACM Comput. Surv. **45**(1), 1–34 (2012)
13. Farina, D., Jiang, N., Rehbaum, H., Holobar, A., Graimann, B., Dietl, H., Aszmann, O.C.: The extraction of neural information from the surface EMG for the control of upper-limb prostheses: emerging avenues and challenges. IEEE Trans. Neural Syst. Rehabil. Eng. **22**(4), 797–809 (2014)

14. Gaddy, D., Klein, D.: Digital voicing of silent speech (2020)
15. Grabocka, J., Schilling, N., Wistuba, M., Schmidt-Thieme, L.: Learning time-series shapelets. In: Proceedings of the 20th ACM SIGKDD International Conference on Knowledge Discovery and Data Mining. ACM (2014)
16. Hahne, J.M., Wilke, M.A., Koppe, M., Farina, D., Schilling, A.F.: Longitudinal case study of regression-based hand prosthesis control in daily life. Front. Neurosci. **14** (2020)
17. Hand, D.J., Yu, K.: Idiot's Bayes: not so stupid after all? Int. Stat. Rev./Revue Internationale de Statistique **69**(3), 385 (2001)
18. Holm, S.: A simple sequentially rejective multiple test procedure. Scand. J. Stat. **6**(2), 65–70 (1979)
19. Hou, W., Li, D., Xu, C., Zhang, H., Li, T.: An advanced k nearest neighbor classification algorithm based on KD-tree. In: 2018 IEEE International Conference of Safety Produce Informatization (IICSPI). IEEE (2018)
20. Hua, X., Li, J., Wang, T., Wang, J., Pi, S., Li, H., Xi, X.: Evaluation of movement functional rehabilitation after stroke: a study via graph theory and corticomuscular coupling as potential biomarker. Math. Biosci. Eng. **20**(6), 10530–10551 (2023)
21. Jabbari, M., Khushaba, R.N., Nazarpour, K.: Combined dynamic time warping and spatiotemporal attention for myoelectric control. In: 2021 43rd Annual International Conference of the IEEE Engineering in Medicine & Biology Society (EMBC). IEEE (2021)
22. Kumar, P., Phinyomark, A., Scheme, E.: Verification-based design of a robust EMG wake word. In: 2021 43rd Annual International Conference of the IEEE Engineering in Medicine & Biology Society (EMBC). IEEE (2021)
23. Kurogi, T., Yamaba, H., Aburada, K., Katayama, T., Park, M., Okazaki, N.: A study on a user identification method using dynamic time warping to realize an authentication system by s-EMG. In: Advances in Internet, Data and Web Technologies, pp. 889–900. Springer, Cham (2018)
24. Kurzynski, M., Krysmann, M., Trajdos, P., Wolczowski, A.: Multiclassifier system with hybrid learning applied to the control of bioprosthetic hand. Comput. Biol. Med. **69**, 286–297 (2016)
25. Kurzynski, M., Trajdos, P., Wolczowski, A.: Multiclassifier system using class and interclass competence of base classifiers applied to the recognition of grasping movements in the control of bioprosthetic hand. In: Progress in Artificial Intelligence, pp. 174–185. Springer, Cham (2017)
26. Parajuli, N., Sreenivasan, N., Bifulco, P., Cesarelli, M., Savino, S., Niola, V., Esposito, D., Hamilton, T.J., Naik, G.R., Gunawardana, U., Gargiulo, G.D.: Real-time EMG based pattern recognition control for hand prostheses: a review on existing methods, challenges and future implementation. Sensors **19**(20), 4596 (2019)
27. Phinyomark, A., Phukpattaranont, P., Limsakul, C.: Feature reduction and selection for EMG signal classification. Expert Syst. Appl. **39**(8), 7420–7431 (2012)
28. Rousseeuw, P.J.: Silhouettes: a graphical aid to the interpretation and validation of cluster analysis. J. Comput. Appl. Math. **20**, 53–65 (1987)
29. Satopaa, V., Albrecht, J., Irwin, D., Raghavan, B.: Finding a "kneedle" in a haystack: detecting knee points in system behavior. In: 2011 31st International Conference on Distributed Computing Systems Workshops. IEEE (2011)
30. Sokolova, M., Lapalme, G.: A systematic analysis of performance measures for classification tasks. Inf. Process. Manag. **45**(4), 427–437 (2009)
31. Tan, S., An, Y., Wu, Y., Zhang, D.: Electromyography based handwriting recognition system using LM-BP neural network. In: 2016 9th International Conference on Human System Interactions (HSI). IEEE (2016)
32. Vijaya, Sharma, S., Batra, N.: Comparative study of single linkage, complete linkage, and ward method of agglomerative clustering. In: 2019 International Conference on Machine Learning, Big Data, Cloud and Parallel Computing (COMITCon). IEEE (2019)
33. Wang, C., Wang, S., Zhao, X., Fang, J., Zhang, M.: EMG-based gesture recognition system and its hardware implementation. In: 2021 2nd International Conference on Big Data & Artificial Intelligence & Software Engineering (ICBASE). IEEE (2021)

34. Wang, Q., Boyer, K.L.: Feature learning by multidimensional scaling and its applications in object recognition. In: 2013 XXVI Conference on Graphics, Patterns and Images. IEEE (2013)
35. Wolczowski, A., Bledowski, M., Witkowski, J.: The system for EMG and MMG singals recording for the bioprosthetic hand control. J. Autom. Mob. Robot. Intell. Syst. **11**(3), 22–29 (2017)
36. Yildiz, K.A., Shin, A.Y., Kaufman, K.R.: Interfaces with the peripheral nervous system for the control of a neuroprosthetic limb: a review. J. Neuroeng. Rehabil. **17**(1) (2020)
37. Yu, M., Li, G., Jiang, D., Jiang, G., Tao, B., Chen, D.: Hand medical monitoring system based on machine learning and optimal EMG feature set. Pers. Ubiquitous Comput. (2019)
38. Zhang, C., Fanaee-T, H., Thoresen, M.: Feature extraction from unequal length heterogeneous EHR time series via dynamic time warping and tensor decomposition. Data Min. Knowl. Disc. **35**(4), 1760–1784 (2021)

A Hybrid Long Short-Term Memory Network Based on Wind and Rain Sensitive and Its Application to PM2.5 Prediction

Bo-Xiang Huang and Shin-Jye Lee

Abstract With the rise of health consciousness in recent years, air pollution has become an issue of public concern. The Air Quality Index (AQI), which takes into account various air pollution factors has become an indispensable part of daily life. Of these factors, PM2.5 has a particularly serious impact on the human body. The final Wind-and-Rainfall sensitive LSTM (WRLSTM) can effectively predict PM2.5 concentrations in the next 6–18 h. The results are better than those of common deep learning algorithms, such as MLP regression.

Keywords Air Pollution Forecasting · Attention Mechanism · Deep Learning · Smart City

1 Introduction

Since the industrial development, the air pollution problem has become more and more serious and has become a global concern. According to the World Health Organization, three of the top ten causes of death in the world are respiratory diseases [1]: third - chronic obstructive pneumonia, fourth - lower respiratory tract infection, and sixth - trachea, bronchus, and lung cancer. Among all pollutants, PM2.5 is the most dangerous factor for health, which can enter the trachea and alveoli through the nose and throat, and may cause serious respiratory diseases [2], and may consequently affect the rate of medical consultations nationally [3] and increase the medical burden [4]. Therefore, the monitoring and control of air pollutants has become an important issue. The Environmental Protection Agency (EPA) in Taiwan has set up 84 central

B.-X. Huang
Department of Computer Science, National Yang Ming Chiao Tung University, Hsinchu 300093, Taiwan

S.-J. Lee (✉)
Institute of Technology Management, National Yang Ming Chiao Tung University, Hsinchu 300093, Taiwan
e-mail: camhero@gmail.com

R. Burduk et al. (eds.), *Progress on Pattern Classification, Image Processing and Communications*, Lecture Notes in Networks and Systems 766,
https://doi.org/10.1007/978-3-031-41630-9_23

air quality monitoring stations to update the concentrations of various pollutants hourly, so that people can check the real-time air pollution information online and decide whether to go out. As the threat of PM2.5 grows, we need a trusted forecasting system to provide early warnings.

Air pollution is influenced by weather and human activities, and varies continuously in time and space [5]. Weather changes, including temperature, humidity, rainfall, wind speed, and wind direction, have a particularly large impact on air pollution [6]. Air pollution prediction is a standard time-series problem. In the field of machine learning and deep learning, recurrent neural networks (RNN), long short-term memory models (LSTM), etc. are commonly used to solve the time series problem. In the past, air pollution prediction was often based solely on historical data of air pollution [7, 8], and multiple meteorological factors were seldom considered together, making it difficult to predict the effect of weather changes on air pollution levels effectively.

This work proposes a two-phase architecture. In the first stage, the LSTM and attention mechanism are used to predict PM2.5 concentrations in neighboring areas by considering the effects of spatial variations of wind direction and rainfall on PM2.5 concentrations in these areas. In the second stage, the aforementioned models are combined, and the attention mechanism is used to consider the effects of wind direction and rainfall simultaneously. The weights of the changes in PM2.5 concentrations are calculated to obtain the final PM2.5 concentration prediction results. The result demonstrate that the proposed method is superior to other machine learning models in predicting PM2.5 concentrations.

2 Related Work

2.1 Air Pollution Forecasting and Air Pollution Factors

Factors affecting the change in air pollution concentration include offshore pollution, weather and other factors. Borge et al. [6] found that temperature and rainfall have a certain degree of influence on the concentration of air pollution sources, and Peng et al. [9] also showed that there are seasonal differences in PM2.5 concentrations, with the highest concentrations in winter and the lowest in summer. Machine learning has been widely used in predicting air pollution, and Lin et al. [10] used three different machine learning algorithms to train and build models. The SVR is a type of regression analysis and differs from linear regression in that it finds an optimal hyperplane to predict the value. XGBoost is a modification and extension of the Gradient Boosted Decision Tree (GBDT) and is used to solve supervised learning problems. It can solve the overfitting problem by adjusting the learning rate, and provides gradient boosting of parallel trees, so it can quickly solve data science problems.

2.2 *Time Series Problem*

By analyzing the trends and cycles of time series data, we can predict future values. Earlier studies utilized recurrent neural networks (RNN) to perform this analysis [11]. LSTM is a type of recurrent neural network that excels at processing and predicting time series data [12]. It is widely used in air pollution prediction, but few studies have examined the effects of wind and rainfall on air pollution at the same time. Unlike traditional RNN, LSTM has four internal layers [13]. The first layer is the forget gate, which determines which information is to be forgotten from the cell state. The second part is the input gate, which determines which new information should be updated into the cell, followed by a layer of tanh function to update the cell. The LSTM controls the information transfer from the previous cell, which can effectively solve the vanishing gradient problem caused by traditional RNNs. Since the air pollution concentration varies continuously over time, LSTM is expected to perform well in this problem.

3 Proposed Approach

3.1 *Overview*

The proposed model comprises four parts: data pre-processing, a wind-sensitive LSTM model, a rainfall-sensitive LSTM model, and a wind-and-rain LSTM model. The model is illustrated in Fig. 1. It employs LSTM and an attention mechanism to consider the effects of spatial variations in wind direction and rainfall on PM2.5 concentrations in neighboring areas, and generates PM2.5 concentration predictions. By integrating the aforementioned models and using the attention mechanism to consider the effects of wind direction and rainfall, we calculate the weights of PM2.5 concentration changes between the models, resulting in the final PM2.5 concentration predictions.

3.2 *Data Preprocessing*

In order to use weather characteristics (especially wind and rainfall) to assist in the time series prediction of PM2.5, it is particularly important to represent the wind characteristics and combine the data.

One-Hot Encoding for Wind Features. In EPA data, wind direction is expressed from 0° to 360°. However, wind directions that are slightly larger than 0° and slightly smaller than 360° are considered opposite directions. To classify wind directions into eight directions $\mathbb{S}$, such as "north", "northeast", "east", and "northwest", we use one-hot encoding to represent them. We encode the categorical variable x_{wd}

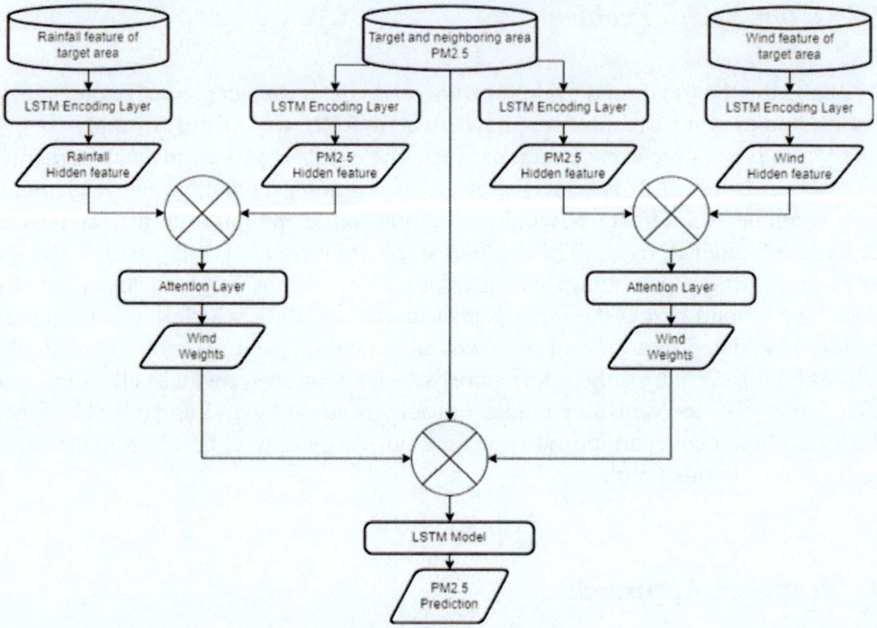

Fig. 1 Overview of the proposed PM2.5 prediction method

into the encoded feature $\overrightarrow{wd}$. Next, we combine the wind direction features $\overrightarrow{wd}$ with wind speed x_{ws} by multiplication into combined wind features $\overrightarrow{wind}$, as shown in Eqs. (1)–(4):

$$f(x, c) = \begin{cases} 1 & \text{if } x = c \\ 0 & \text{if } x \neq c \end{cases}, c \in \mathbb{S}. \tag{1}$$

$$f_{onehot}(x, \mathbb{S}) = \langle f(x, c_1), f(x, c_2), \ldots, f(x, c_s) \rangle \tag{2}$$

$$\overrightarrow{wd} = f_{onehot(x_{wd}, \mathbb{S})} \tag{3}$$

$$\overrightarrow{wind} = \overrightarrow{wd} \times X_{ws} \tag{4}$$

Making Time Series Features

Wind Time Series. We consider the number of m past time period for wind at the target area by current timestamp ct, when generating time series data of wind $\overrightarrow{Wind_{ct}}$, as shown in Eq. (5):

$$\overrightarrow{Wind_{ct}} = (\overrightarrow{Wind_{ct-m+1}}, \overrightarrow{Wind_{ct-m+2}}, \ldots, \overrightarrow{Wind_{ct}}) \tag{5}$$

Rainfall Time Series Features. We consider the number of m past time period for rainfall by current timestamp ct and the data of neighboring area $\mathbb{L}$ are joined to generate time series data of rainfall $\overrightarrow{Rain_{l,ct}}$, as shown in Eq. (6):

$$\overrightarrow{Rain_{L,ct}} = \begin{pmatrix} Rain_{1,ct-m+1} & \cdots & Rain_{l,ct-m+1} & \cdots & Rain_{L,ct-m+1} \\ \vdots & \ddots & \vdots & & \vdots \\ Rain_{1,ct-m+n} & \cdots & Rain_{l,ct-m+n} & \cdots & Rain_{L,ct-m+n} \\ \vdots & & \vdots & \ddots & \vdots \\ Rain_{1,ct} & \cdots & Rain_{l,ct} & \cdots & Rain_{L,ct} \end{pmatrix} \tag{6}$$

where $l \in \mathbb{L}$, $1 < n < m$.

PM2.5 Time Series Features. Similar to the rainfall time series, we consider the number of m past time period for rainfall by current timestamp ct and take the data of neighboring area $\mathbb{L}$ into consideration as shown in Eq. (7):

$$\overrightarrow{PM2.5_{L,ct}} = \begin{pmatrix} PM2.5_{1,ct-m+1} & \cdots & PM2.5_{l,ct-m+1} & \cdots & PM2.5_{L,ct-m+1} \\ \vdots & \ddots & \vdots & & \vdots \\ PM2.5_{1,ct-m+n} & \cdots & PM2.5_{l,ct-m+n} & \cdots & PM2.5_{L,ct-m+n} \\ \vdots & & \vdots & \ddots & \vdots \\ PM2.5_{1,ct} & \cdots & PM2.5_{l,ct} & \cdots & PM2.5_{L,ct} \end{pmatrix} \tag{7}$$

where $l \in \mathbb{L}$, $1 < n < m$.

3.3 *Wind-Sensitive LSTM and Rain-Sensitive LSTM*

Using the core concept of attention mechanism, we focus on the effect of wind or rainfall on PM2.5 concentration and use the obtained feature weights in the prediction of LSTM model. The model is illustrated in Fig. 2. The RSLSTM has same architecture.

4 Experiment and Evaluation

This section presents the data set and experiment setup. Evaluations of all compared models based on real weather forecasts are presented.

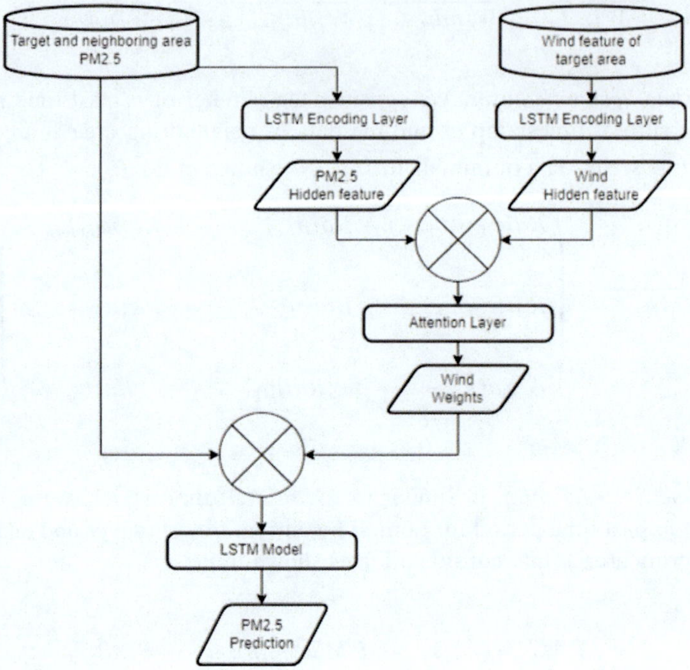

Fig. 2 The architecture of WSLSTM

4.1 Dataset and Experiment Setup

This work uses the historical PM2.5 monitoring data provided by the Environmental Protection Administration of the Executive Yuan in Taiwan as the dataset. We take Wanhua as target area, while the surrounding concentrations were measured at Zhongshan, Guting, Yonghe, Banqiao, Xinzhuang, Cailiao, Songshan, and Shilin stations as the data sources. The data from 2018/01/01 to 2020/12/31 was used as the training set and 2021/01/01 to 2021/12/31 was used as the test set.

4.2 Evaluation Metric

This work compares the predictive accuracy of different parameter combinations or algorithms through mean absolute error (MAE) and root-mean-square error (RMSE) evaluation metrics, which are widely used to measure the residuals between predicted values and those observed in regression analysis. As defined in Eq. (8) and (9):

$$MAE = \frac{1}{n}\sum_{i=1}^{n}|\widehat{y_i} - y_i| \tag{8}$$

$$RMSE = \sqrt{\frac{1}{n}\sum_{i=1}^{n}(\widehat{y_i} - y_i)^2} \quad (9)$$

where $\widehat{y_i}$ is the predicted PM2.5 concentration, y_i corresponds to the observed PM2.5 concentration, and n is the number of test samples.

4.3 Compared Models

This experiment compares the predictive accuracy of various parameter combinations and algorithms using two widely-used evaluation metrics: root-mean-square error (RMSE) and mean absolute error (MAE). To provide a basis for comparison, we include several common models in our analysis:

- Linear Regression (LR): A model that uses a least squares function called a linear regression equation to model the relationship between one or more independent variables and the dependent variable.
- Multilayer Perceptron (MLP): A type of forward-structured artificial neural network that maps input vectors to output vectors.
- eXtreme Gradient Boosting (XGboost): A method that uses gradient boosting trees to solve data science problems quickly and accurately.
- Long Short-Term Memory Network (LSTM): A recurrent neural network (RNN) that can effectively model dynamic time series behavior.
- Wind-Sensitive LSTM (WSLSTM): A variant of LSTM that includes an attention mechanism to improve the modeling of wind direction and speed.
- Rain-Sensitive LSTM (RSLSTM): An improved version of LSTM that incorporates an attention mechanism to better model rainfall.
- Wind-and-Rain-Sensitive LSTM (WRLSTM): A long- and short-term memory model that incorporates an attention mechanism to improve the modeling of wind speed, wind direction, and rainfall.

4.4 Evaluation of the Number of Past Time Periods for PM2.5

The historical data of 4, 8 and 12 h were used to predict the PM2.5 concentrations for the next 6, 12 and 18 h. It was found that the longer historical data did not reduce the MAE and RMSE, which is presumed to be due to the temporality of rainfall and wind direction. The result is shown in Fig. 3 and Fig. 4.

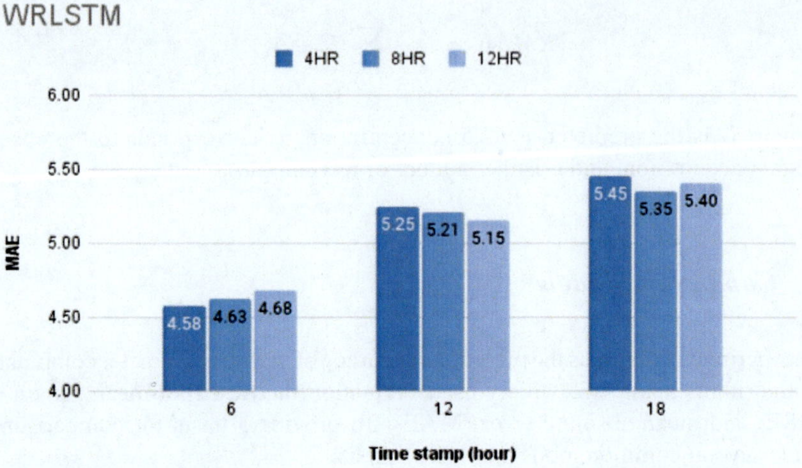

Fig. 3 Evaluation of different past time period with MAE

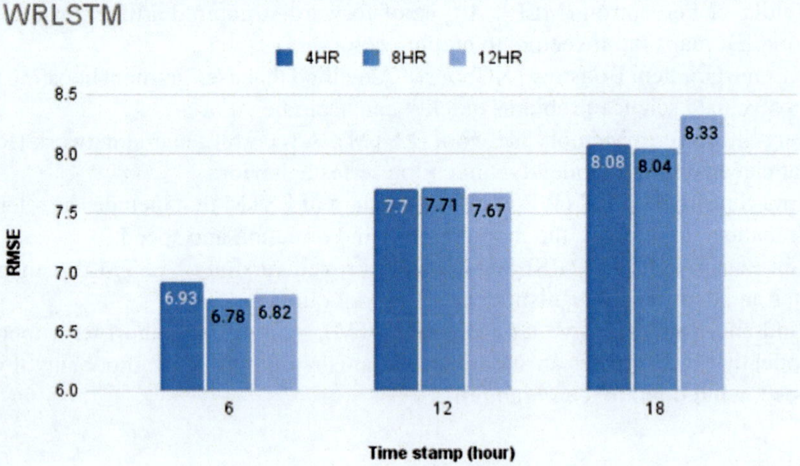

Fig. 4 Evaluation of different past time period with RMSE

Table 1 Comparison of all models with MAE

Hour	MLP	XGBoost	Linear Regression	LSTM	WSLSTM	RSLSTM	WRLSTM
6	4.9743	4.8411	4.9533	4.7372	**4.6374**	4.7219	4.6772
12	5.5166	5.4193	5.5468	5.5198	5.3133	5.2454	**5.1876**
18	5.8393	5.613	5.7297	5.705	5.4675	5.4731	**5.4443**

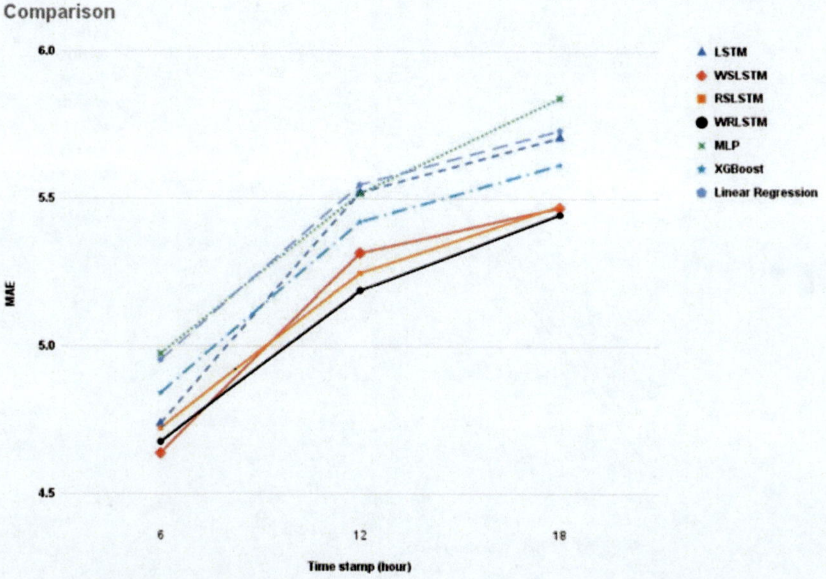

Fig. 5 Comparison of all models with MAE

4.5 *Evaluation of All Comparison Models*

WRLSTM outperforms other common algorithms in predicting PM2.5 concentrations in the next 6, 12, and 18 h. The result with MAE is shown in Table 1 and Fig. 5. The result with RMSE is shown in Table 2 and Fig. 6.

Table 2 Comparison of all models with RMSE

Hour	MLP	XGBoost	Linear Regression	LSTM	WSLSTM	RSLSTM	WRLSTM
6	7.11	7.09	7.03	6.93	6.9	6.85	**6.78**
12	7.83	7.94	7.87	7.8	7.87	7.73	**7.71**
18	8.1753	8.12	8.1415	8.11	8.1075	8.1021	**8.04**

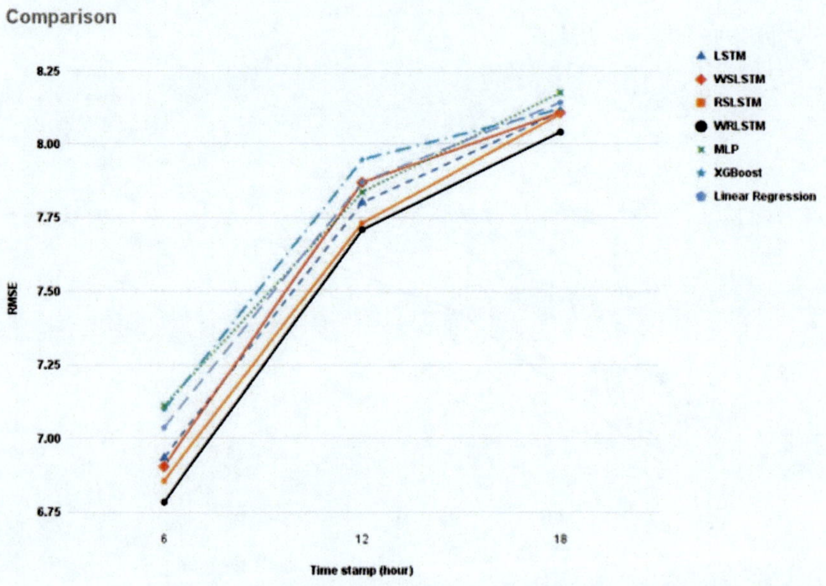

Fig. 6 Comparison of all models with RMSE

Acknowledgements This research was funded by Taiwan National Science and Technology Council, grant number MOST 111-2410-H-A49-019-.

References

1. The top 10 causes of death. WHO. https://www.who.int/news-room/fact-sheets/detail/the-top-10-causes-of-death. Accessed 9 Dec 2020
2. Xing, Y.F., Xu, Y.H., Shi, M.H., Lian, Y.X.: The impact of PM2.5 on the human respiratory system. J. Thorac. Dis. **8**(1), E69–E74 (2016). https://doi.org/10.3978/j.issn.2072-1439.2016.01.19
3. Huang, Y.-J.: A panel data analysis of air pollution and hospital outpatients and admissions for respiratory and cardiovascular diseases in Taiwan. National Digital Library of Theses and

Dissertations in Taiwan (2015). https://hdl.handle.net/11296/zeepxz
4. Yu, C.-H.: A study on the correlation between air pollutants, meteorological factors, and hospital admissions of respiratory diseases - Using Taipei City as an example. National Digital Library of Theses and Dissertations in Taiwan (2009). https://hdl.handle.net/11296/w4vqus
5. Huang, P., Zhang, J., Tang, Y., Liu, L.: Spatial and temporal distribution of PM2.5 pollution in Xi'an City, China. Int. J. Environ. Res. Public Health **12**(6), 6608–6625 (2015). https://doi.org/10.3390/ijerph120606608
6. Borge, R., Requia, W. J., Yagüe, C., Jhun, I., Koutrakis, P.: Impact of weather changes on air quality and related mortality in Spain over a 25-year period [1993-2017]. Environ. Int. **133**(Pt B), 105272 (2019). https://doi.org/10.1016/j.envint.2019.105272
7. Huang, W.-S.: The Prediction for Time-Series Data - a Case Study of Air Pollution Open Data at Taichung City. National Digital Library of Theses and Dissertations in Taiwan (2017). https://hdl.handle.net/11296/n9cutn
8. Yang, H.-Y.: The correlation of air quality and weather patterns in Taiwan territory. National Digital Library of Theses and Dissertations in Taiwan (1993). https://hdl.handle.net/11296/mn8qup
9. Peng, R.D., Dominici, F., Pastor-Barriuso, R., Zeger, S.L., Samet, J.M.: Seasonal analyses of air pollution and mortality in 100 US cities. Am. J. Epidemiol. **161**(6), 585–594 (2005). https://doi.org/10.1093/aje/kwi075
10. Lin, K.-M.: Air Pollution Forecasting Using Machine Learning Methods on Big Data Platform. National Digital Library of Theses and Dissertations in Taiwan (2018). https://hdl.handle.net/11296/dvxvgw
11. Zhao, X., Zhang, R., Wu, J., Chang, P.C.: A deep recurrent neural network for air quality classification. J. Inf. Hiding Multim. Signal Process. **9**, 346–354 (2018)
12. Xayasouk, T., Lee, H., Lee, G.: Air pollution prediction using long short-term memory (LSTM) and deep autoencoder (DAE) models. Sustainability **12**(6), 2570 (2020). https://doi.org/10.3390/su12062570
13. Hochreiter, S., Schmidhuber, J.: Long short-term memory. Neural Comput. **9**(8), 1735–1780 (1997)

Author Index

R. Burduk et al. (eds.), *Progress on Pattern Classification, Image Processing and Communications*, Lecture Notes in Networks and Systems 766,
https://doi.org/10.1007/978-3-031-41630-9